STUDENT'S SOLUTIONS MA
UNDERST
CHEMISTRY

FRED M. DEWEY

PREPARED BY
JULIETTE A. BRYSON
LAS POSITAS COLLEGE

WEST PUBLISHING COMPANY
Minneapolis/St. Paul   New York   Los Angeles   San Francisco

WEST'S COMMITMENT TO THE ENVIRONMENT

In 1906, West Publishing Company began recycling materials left over from the production of books. This began a tradition of efficient and responsible use of resources. Today, up to 95% of our legal books and 70% of our college texts and school texts are printed on recycled, acid-free stock. West also recycles nearly 22 million pounds of scrap paper annually—the equivalent of 181,717 trees. Since the 1960s, West has devised ways to capture and recycle waste inks, solvents, oils, and vapors created in the printing process. We also recycle plastics of all kinds, wood, glass, corrugated cardboard, and batteries, and have eliminated the use of Styrofoam book packaging. We at West are proud of the longevity and the scope of our commitment to the environment.

Production, Prepress, Printing and Binding by West Publishing Company.

COPYRIGHT © 1994 by   WEST PUBLISHING CO.
          610 Opperman Drive
          P.O. Box 64526
          St. Paul, MN 55164–0526

All rights reserved
Printed in the United States of America
01 00 99 98 97 96 95 94 8 7 6 5 4 3 2 1 0

ISBN 0–314–03668–7

CONTENTS

Chapter 1	Chemistry—The Central Science	1
Chapter 2	Matter and Energy	3
Chapter 3	Measurements	9
Chapter 4	Problem Solving	17
Chapter 5	The Structure of Atoms. The Periodic Table	31
Chapter 6	Electron Structure and the Periodic Table	39
Chapter 7	Composition and Formulas of Compounds	47
Chapter 8	The Structure of Compounds. Chemical Bonds	69
Chapter 9	Names and Formulas of Inorganic Compounds	79
Chapter 10	Chemical Equations	87
Chapter 11	Calculations Involving Chemical Equations	109
Chapter 12	The Gaseous State	127
Chapter 13	Liquids, Solids, and Changes of State	147
Chapter 14	Solutions	157
Chapter 15	Acids and Bases	177
Chapter 16	Reactions Rates and Chemical Equilibrium	185
Chapter 17	Oxidation–Reduction Reactions	193
Chapter 18	Radioactivity and Nuclear Energy	207
Chapter 19	Introduction to Organic Chemistry	213
Chapter 20	Biochemistry: The Chemistry of Life	219

Chapter 1:
Chemistry—The Central Science

1.1 **(b)** sunlight **(d)** electricity **(g)** heat **(h)** love **(i)** fear

Matter is anything that has mass and occupies space. The items above do not meet this definition and are therefore not classified as matter.

1.3 An analytical chemist could provide information to the physician about the quantity of a substance in a patient's blood.

1.5 The division of environmental chemistry would be most involved with air and water pollution studies. Much of the work of environmental chemists involves analytical chemistry, since it is important to determine the identity and quantity of pollutants.

1.6 organic: **(a)** sugar from sugar cane   inorganic: **(b)** table salt
        (c) a piece of wood   **(e)** sand
        (d) human hair   **(f)** water

Organic chemistry is the study of compounds of carbon, and most living matter is based on carbon. Inorganic chemistry is the study of materials primarily of mineral origin.

1.10 A precise statement of a problem is important to a scientist so that he or she has a clear focus in choosing a direction to follow in solving a problem.

1.11 (a) hypothesis—a tentative answer
    (b) law—a statement of natural events
    (c) law—a statement of natural events
    (d) theory—an explanation supported by significant amount of data
    (e) hypothesis—a tentative answer
    (f) theory—an explanation supported by significant amount of data
    (g) hypothesis—a tentative answer

Chapter 2:
Matter and Energy

2.1 solid—definite shape and definite volume

liquid—indefinite shape but definite volume

gas—indefinite shape and indefinite volume

2.3 The kinetic theory of matter describes matter as composed of tiny particles that are in constant random motion. The motion of the particles increases with temperature. Thus, there is a little motion of the particles in a solid—mostly vibrational. As a solid is heated, the particles acquire more energy. At some point there is enough energy for the particles to break away from one another and slide around. This is the liquid state. As heat is added and the temperature continues to increase, the particles have enough energy to completely escape from one another and the substance enters the gaseous state.

2.5 (**b**) air (**c**) motor oil (**e**) gasoline

All of these substances can flow; therefore, they are all fluids.

2.7 homogeneous: (**a**) salt water (**e**) gasoline (**f**) vegetable oil heterogeneous: **(b)** Italian salad dressing **(c)** iced tea (assuming there are actual ice cubes in the tea) **(d)** orange juice **(g)** grape jam **(h)** concrete

A homogeneous mixture is uniform in composition, properties, and appearance. Heterogeneous mixtures are non-uniform.

2.9 **(a)** hydrogen **(c)** aluminum **(h)** oxygen

Common elements are listed in Table 2.1.

2.11 mixture

The remaining substance was clearly different. It had no odor and was not flammable. The flammable alcohol burned, leaving behind nonflammable water.

2.13 single substance

Single substances have definite composition, while mixtures have variable composition.

2.15 **(a)** N is nitrogen **(b)** P is phosphorus **(c)** Si is silicon **(d)** Li is lithium
(e) Cl is chlorine **(f)** He is helium

2.17 **(a)** 2 **(b)** 3 **(c)** 6

The subscript following each element in a formula gives the number of atoms of that element in each molecule. Since CO_2 has the subscript 2 after the symbol for oxygen there are 2 atoms of oxygen per molecule of CO_2. Similarly the subscript 3 in SO_3 and the subscript 6 in $C_6H_{12}O_6$ identify the number of oxygen atoms.

2.19 (a) NaH   sodium hydride   (b) CCl_4, carbon tetrachloride   (c) CaS, calcium sulfide

2.21 (a) Mg_3N_2, magnesium nitride   (b) LiCl, lithium chloride   (c) Al_2O_3, aluminum oxide

2.23 (b) exploding a firecracker   (g) lighting a fire

New substances with different properties are formed in chemical changes.

2.25 mixture; no chemical change; yes, a chemical change occurs; water, H_2O

Simply mixing the gases does not produce new substances; therefore, the result is a mixture and no chemical change has taken place. The formation of a liquid after ignition of the mixture indicates that a chemical change has taken place. The liquid is water, which has the formula H_2O.

2.27 (b), (c), and (e) are not balanced.

(b) $Li + O_2 \rightarrow Li_2O$ Neither the lithium nor the oxygen atoms are balanced. There are two lithium atoms on the product side and only one lithium on the reactant side. There are two oxygen atoms on the left side and only one oxygen atom on the right side.

(c) $CH_4 + O_2 \rightarrow CO_2 + H_2O$ There are only 2 oxygen atoms on the left side of the equation, whereas there are four oxygen atoms on the right side of the equation.

(e) $N_2 + O_2 \rightarrow 2\ NO_2$ The oxygen atoms are not balanced. There are two oxygen atoms on the left side of the equation and four oxygen atoms on the right side of the equation.

2.29   (b), (c), and (e)

To be separable by filtration, there must be solid particles that can be retained by filter paper. Noodles, sawdust, and broken glass are solids.

2.31   chromatography

Chromatography is based upon the differing attractions of substances for two media, one of which is stationary (such as paper) and one of which is mobile (such as water).

2.33   positive

Since opposite charges are attracted to one another, the negative Cl^- ions would be attracted toward a positive charge.

2.35   exothermic: (a)   A burning candle releases energy in the form of light and heat.
                    (e)   A burning marshmallow releases energy in the form of light and heat.
       endothermic: (b)   Pancakes cooking on a griddle require heat input to keep cooking.

(c) Digesting a meal consumes energy to break complex foods into simpler substances.

(d) A roasting marshmallow is consuming heat as it is converted to a tastier form.

2.37 potential energy decreases; kinetic energy increases; no change in total energy

Potential energy decreases as the apple falls out of the tree, since potential energy depends on position. As it falls, the stored or potential energy of the apple is converted into kinetic energy. Kinetic energy is the energy of motion and it increases as the apple falls. Since energy is conserved, there is no change in the total energy of the apple when it falls.

2.39 Inverting a test tube and placing it in water traps a clear colorless gas. Air resists being compressed by the water as you push down on the test tube. Since the air takes up space, it must be matter.

2.41 chemical: (a) and (c)   physical: (b)

(a) The fact that cooking oil burns is a chemical property, because in the process of burning, new substances with different properties are formed.

(b) Gold is yellow in color is a physical property because color can be observed without changing the composition of gold.

(c) A black coating forms on copper when it is heated in air is indicative of a chemical property, since the black coating has different properties (i.e., a different color) from copper.

2.43 (a) CaF_2, calcium fluoride, (b) N_2O, dinitrogen oxide, (c) XeF_4, xenon tetrafluoride

2.45 filtration: **(a)**, **(c)**, and **(d)**   distillation: **(b)**

Filtration can be used to separate mixtures of solids in liquids, thus removing carrots, ice, and cereal from the liquids with which they are mixed. Since the paraffin wax formed a solution, or homogeneous mixture, with the gasoline, filtration will not work and distillation must be used.

Chapter 3:
Measurements

3.1   qualitative: (**a**) and (**b**); quantitative: (**c**)

Quantitative measurements are expressed numerically and result from measurements as in answer (c) a ten-story building. Qualitative observations are not numerical as in (a) red hair and (b) a long rope.

3.3   (**a**) m (meter)   (**b**) s (second)   (**c**) K (kelvin)   (**d**) kg (kilogram)

  (a)   The SI unit for distance is the meter.
  (b)   The SI unit for time is the second.
  (c)   The SI unit for temperature is the kelvin.
  (d)   The SI unit for mass (quantity of tomatoes) is the kilogram.

3.5   (**a**) 1 000 000 m = 1 Mm   (**b**) 0.001 m = 1 mm   (**c**) 100 m = 1 hm   (**d**) 0.000 000 001 m = 1 nm
(**e**) 0.01 m = 1 cm   (**f**) 0.000 001 m = 1 μm

3.7 (a) 3.45×10^{-3} (b) 1.032×10^{2} (c) 4.5623×10^{4}

(a) Since the decimal place must be moved three places to the right, the exponent of ten is -3.

(b) Since the decimal place must be moved two places to the left, the exponent of ten is 2.

(c) Since the decimal place must be moved four places to the left, the exponent of ten is 4.

3.9 602 000 000 000 000 000 000 000

6.02 E 23

The decimal point must be moved 23 places to the right to get back to decimal form.

3.11 (a) 5.40×10^{-2} (rounded to three significant figures) (b) 9.45×10^{12}

(a) 0.732×10^{-2} (b) 0.78×10^{12}
      4.67×10^{-2} 8.67×10^{12}
      ─────────   ─────────
      5.40×10^{-2} 9.45×10^{12}

3.13 (a) 3.47×10^{12} (b) 2.41×10^{10} (c) 9.76×10^{-8}

(a) $(4.74 \times 10^{4})(7.32 \times 10^{7}) = (4.74 \times 7.32)(10^{4} \times 10^{7}) = 34.7 \times 10^{11} = 3.47 \times 10^{12}$

(b) $(3.66 \times 10^{3})(6.59 \times 10^{6}) = (3.66 \times 6.59)(10^{3} \times 10^{6}) = 24.1 \times 10^{9} = 2.41 \times 10^{10}$

(c) $(9.12 \times 10^{-3})(1.07 \times 10^{-5}) = (9.12 \times 1.07)(10^{-3} \times 10^{-5}) = 9.76 \times 10^{-8}$

3.15 (a) 9.25 (b) 1.03×10^{-4}

(a) $\dfrac{(1.25 \times 10^8)(3.78 \times 10^{-3})}{5.11 \times 10^4} = 9.25$ (b) $\dfrac{8.73 \times 10^{-5}}{(2.31 \times 10^{-7})(3.68 \times 10^6)} = 1.03 \times 10^{-4}$

3.17 Mass expresses the quantity of matter in a sample. Mass is independent of location and/or gravity. Weight measures the force exerted by a mass under the influence of gravity. The mass of an astronaut's suit will not change when taken from the earth to the moon, but the weight of the suit will be much less on the moon because the force of gravity is less on the surface of the moon than it is on the surface of the earth.

3.19 (a) 1 000 000 g (b) 0.001 g (c) 0.01 g (d) 10 g (e) 0.000 000 001 g (f) 1000 g

(a) M stands for mega, which equals 10^6 or 1 000 000. (b) m stands for milli, which equals 10^{-3} or 0.001. (c) c stands for centi, which equals 10^{-2}. (d) da stands for deka, which equals 10. (e) n stands for nano, which equals 10^{-9} or 0.000 000 001. (f) k stands for kilo, which equals 10^3 or 1000.

3.21 "Two dimensional" quantities are those requiring 2 dimensions in their calculation. Area is an example of a two dimensional quantity, as the area of an office could be 2.5 m by 3.0 m = 7.5 m². "Three dimensional" quantities are those requiring 3 dimensions in their calculation. Volume is the most common example of a three dimensional quantity: the volume of a box is calculated 25.0 cm x 25.0 cm x 20.0 cm = 1.25×10^4 cm³.

3.23 (a) a pipet

Pipets are very precisely calibrated glassware. Beakers are suitable only for estimations, and graduated cylinders are only graduated in milliliters.

3.25 liquids – g/mL or g/cm^3; solids – g/cm^3; gases – g/L or g/mL

Densities for liquids and solids are of the same order of magnitude, so the same units are commonly used in their measurement. Remember that the volume occupied by 1 mL is the same as the volume occupied by 1 cm^3. Gases are much less dense than liquids. If gas densities were reported in g/mL, the values would be small. Instead, gas densities are commonly expressed in g/L. See Table 3.4 in the textbook.

3.27 The density of water (0.998 g/mL or 998 g/L) is about 1000 times greater than the density of air (1.2 g/L). Fluid must be pushed out of the way when running through it. Since a volume of water is more massive than an equal volume of air, more work must be done (more exertion) when running in water than when running in air.

3.29 (a) A calorie is a quantity of energy, specifically the amount of energy required to raise the temperature of 1.0 gram of water by one degree Celsius. (b) A food calorie is also a quantity of energy. It is the unit used by nutritionists to describe the energy value of foods. One Calorie contains the same amount of energy as one kilocalorie or 1000 calories. (c) Absolute zero is the theoretically lowest temperature possible. It is equal to -273.15°C or 0 K.

3.31 (a) a pound of gold   (b) a gallon of water   (c) ten bricks

The quantity of heat released depends on three things: mass, specific heat, and the temperature change. In each of the questions above, the specific heat was the same and the temperature change was the same. The only variable was the mass. The object with the greater mass releases more heat to the surroundings when it cools down.

3.33 K (kelvin) is the SI unit for temperature.

3.35 (a) $T_F = 1.80 T_C + 32$ or $T_C = \dfrac{T_F - 32}{1.80}$ (b) $T_K = T_C + 273$ or $T_C = T_K - 273$

3.37 (a) 200 K   (b) 90°C

(a) 200 K = -73°C. Since a negative Celsius temperature is lower than 100°F, you can determine that 200 K is less than 100°F. For further proof, $100°F = \dfrac{100°F - 32°F}{1.80} = 38°C.$ -73°C < 38°C.

(b) Since 212°F is the boiling point of water, it is equal to 100°C. 90°C < 100°C.

3.39 Numbers from measurements differ from exact numbers in that measured numbers always contain uncertainty. The last digit recorded is an estimated value. There are no estimates in exact numbers.

3.41 The accuracy of a measurement refers to how well the measurement agrees with the true value. Since precision refers to the repeatability of a measurement without reference to accuracy, it is possible to have measurements with good precision and poor accuracy. This condition occurs when the measuring device is incorrectly calibrated, but correctly used.

3.43 The estimated digit is determined by judging the distance between the smallest calibration marks on the measuring device. Typically, it is first determined whether the measurement is more than half way, about half way, or less than half way between the calibrations marks. If more than half way, how much more? In this way, an estimated value for the last digit is obtained. If it is estimated that the measurement falls on a calibration mark, this is reported by estimating a zero as the last digit.

3.45 (c) a zero between two nonzero digits

Don't be fooled by item (b). It is possible to have zeros to the right of the decimal point that are not significant. Consider the number 0.004 56. There are two zeros after the decimal point that are nonetheless leading zeros.

3.47 (a) 3   (b) 2   (c) 1   (d) 2   (e) 4   (f) 3

3.49 (a) 0.102   (b) 10.8   (c) 213   (d) 1.34×10^4

3.51 (a) 3.61×10^1 (b) 2.03×10^5 (c) 4.40×10^{-4} (d) 7.65×10^2

3.53 (a) 79 m^2 (b) 1.62 g   (c) 57 mi/hr

  (a) multiplication; both numbers have 2 SF (significant figures)
  (b) multiplication; 5 is an exact number, 0.324 g has 3 SF.
  c) division; 9.5 hr has only 2 SF, therefore the answer can have only 2 SF.

3.55 5.00 g

10 tablets x 0.500 g per tablet = 5.00 grams; 10 is an exact number, therefore the answer may have 3 SF.

3.57 181 lb

$$\begin{array}{r} 123 \text{ lb} \\ 34.5 \text{ lb} \\ 5.8 \text{ lb} \\ \underline{18 \quad \text{ lb}} \\ 181.3 \text{ lb} \end{array}$$

The answer must be rounded to agree with the place value of 123 lb and 18 lb.

3.59 0.095 oz per person per day

Weight of soap used = 4.75 oz - 2.1 oz = 2.7 oz

Weight of soap used per day = $\dfrac{2.7 \text{ oz}}{14 \text{ days}}$ = 0.19 oz/day

Weight of soap used per day per person = $\dfrac{0.19 \text{ oz/day}}{2 \text{ persons}}$ = 0.095 oz/day/person

3.61 Fahrenheit thermometer

Since it is desirable for the meat to be solidly frozen, a temperature below the freezing point of water is needed. 0°C is at the freezing point of water, and water coexists with ice at this temperature. 0°F is equal to $\dfrac{0°F - 32°F}{1.80}$ = -18°C and is therefore colder than the freezing point of water. The meat will be solidly frozen at this temperature.

3.63 (a) 4 SF's (b) 3 SF's (c) 4 SF's (d) exact number

3.65 (a) 16.0 mL (b) 36.5°C (c) 12.00 cm

(c) The 12 centimeter mark is the 120 millimeter mark. To indicate that the estimated digit is on the line, one more zero is needed: 120.0 mm.

Chapter 4:
Problem Solving

4.1   basic math and basic algebra

While many problems can be solved with the four basic mathematical functions of addition, subraction, multiplication and division, other problems will require the use of basic algebra. A handheld scientific calculator facilitates solving problems.

4.3   (1) help in getting started and (2) aid to checking the solution.

Outlining a problem can help you get started on a solution. An outline is also an aid to checking your work on completion to see if it is reasonable.

4.5   dimensional analysis and algebra

The two methods used to solve most problems are dimensional analysis and algebra.

4.7   a ratio that expresses the relationship of one quantity to another quantity

Conversion factors are ratios that correspond to equivalence statements. Multiplying a quantity by a conversion factor does not change the value, just the units in which it is expressed.

4.9 (a) $\dfrac{1\text{ lb}}{16\text{ oz}}\quad \dfrac{16\text{ oz}}{1\text{ lb}}$ (d) $\dfrac{1\text{ tsp}}{5\text{ mL}}\quad \dfrac{5\text{ mL}}{1\text{ tsp}}$

(b) $\dfrac{2.20\text{ lb}}{1\text{ kg}}\quad \dfrac{1\text{ kg}}{2.20\text{ lb}}$ (e) $\dfrac{1760\text{ yd}}{1\text{ mi}}\quad \dfrac{1\text{ mi}}{1760\text{ yd}}$

(c) $\dfrac{10^6\,\mu\text{g}}{1\text{ g}}\quad \dfrac{1\text{ g}}{10^6\,\mu\text{g}}$

4.11 1.5×10^6 mm

Find: d (mm) = ?

Given: 1500 m

Known: 1 m = 1000 mm, or $\dfrac{1\text{ m}}{1000\text{ mm}}$, or $\dfrac{1000\text{ mm}}{1\text{ m}}$

Solution: d (mm) = $(\dfrac{1000\text{ mm}}{1\text{ m}})(1500\text{ m}) = 1.5 \times 10^6$ mm

4.13 1.19×10^3 g

Find: m (g) = ?

Given: 2 lb 10 oz

1 lb = 454 g or $\dfrac{454\text{ g}}{1\text{ lb}}$, 1 lb = 16 oz or $\dfrac{1\text{ lb}}{16\text{ oz}}$

Solution: m (g) = mass in grams from pounds plus mass in grams from ounces

(1) mass in pounds from ounces

m (lb) = $(\dfrac{1\text{ lb}}{16\text{ oz}})(10\text{ oz}) = 0.63$ lb

(2) total mass in pounds

2 lb (exact number) + 0.63 lb = 2.63 lb

m (g) = $(\dfrac{454\text{ g}}{1\text{ lb}})(2.63\text{ lb}) = 1.19 \times 10^3$ g

Chapter 4: Problem Solving   19

4.15   3.6 g/cup

Find: $\dfrac{m\ (g)}{cup} = ?$

Given: 1 pkg = 50 cups or $\dfrac{1\ pkg}{50\ cups}$, 1 pkg = 6.4 oz or $\dfrac{1\ pkg}{6.4\ oz}$,

1 lb = 454 g or $\dfrac{1\ lb}{454\ g}$, 1 lb = 16 oz or $\dfrac{1\ lb}{16\ oz}$

Solution: $\dfrac{m\ (g)}{cup} = (\dfrac{454\ g}{1\ \cancel{lb}})(\dfrac{1\ \cancel{lb}}{16\ \cancel{oz}})(\dfrac{6.4\ \cancel{oz}}{1\ \cancel{pkg}})(\dfrac{1\ \cancel{pkg}}{50\ cups}) = \dfrac{3.6\ g}{cup}$

4.17 6.8×10^3 g

Find: m (g) = ?

Given: 1 penny = 3.4 g or $\dfrac{1\ penny}{3.4\ g}$ and $20.00 of pennies (exact)

Known: $1.00 = 100 pennies or $\dfrac{\$1.00}{100\ pennies}$

Solution: m (g) = $(\dfrac{3.4\ g}{1\ \cancel{penny}})(\dfrac{100\ \cancel{pennies}}{\cancel{\$1.00}})(\$20.00) = 6.8 \times 10^3$ g

4.19   45 L

Find: V (L) = ?

Given: 12 gal, 1.000 L = 1.0567 qt or $\dfrac{1.000\ L}{1.0567\ qt}$

Known: 1 gal = 4 qt or $\dfrac{1\ gal}{4\ qt}$

Solution: V (L) = $(\dfrac{1.000\ L}{1.0567\ \cancel{qt}})(\dfrac{4\ \cancel{qt}}{1\ \cancel{gal}})(12\ \cancel{gal}) = 45$ L

4.21   0.75 L/pkg

Find: liters per package $(\dfrac{L}{pkg}) = ?$

Given: 1 package = 3 boxes or $\frac{1\text{ pkg}}{3\text{ box}}$, 1 box = 8.5 oz or $\frac{1\text{ box}}{8.5\text{ oz}}$, 1 qt = 32 oz or $\frac{1\text{ qt}}{32\text{ oz}}$

Known: 1.000 L = 1.0567 qt or $\frac{1.000\text{ L}}{1.0567\text{ qt}}$

Solution: $V\left(\frac{L}{\text{pkg}}\right) = \left(\frac{1.000\text{ L}}{1.0567\text{ qt}}\right)\left(\frac{1\text{ qt}}{32\text{ oz}}\right)\left(\frac{8.5\text{ oz}}{1\text{ box}}\right)\left(\frac{3\text{ box}}{1\text{ pkg}}\right) = \frac{0.75\text{ L}}{\text{pkg}}$

4.23 25 doses

Find: doses = ?

Given: 250 mL, 1 tsp = 5 mL or $\frac{1\text{ tsp}}{5\text{ mL}}$, 1 dose = 2 tsp or $\frac{1\text{ dose}}{2\text{ tsp}}$

Solution: doses = $\left(\frac{1\text{ dose}}{2\text{ tsp}}\right)\left(\frac{1\text{ tsp}}{5\text{ mL}}\right)(250\text{ mL}) = 25$ doses

4.25 3×10^1 cups

Find: # cups = ?

Given: 1 cup = 8 oz or $\frac{1\text{ cup}}{8\text{ oz}}$, 1 qt = 32 oz or $\frac{1\text{ qt}}{32\text{ oz}}$, 1 gal = 4 qt or $\frac{1\text{ gal}}{4\text{ qt}}$, 2 gal

Solution: # cups = $\left(\frac{1\text{ cup}}{8\text{ oz}}\right)\left(\frac{32\text{ oz}}{1\text{ qt}}\right)\left(\frac{4\text{ qt}}{1\text{ gal}}\right)(2\text{ gal}) = 3 \times 10^1$ cups

Note: The numbers 32 and 4 are exact numbers, but the numbers 8 and 2 each have only one significant figure. Therefore the answer may have only one significant figure.

4.27 $\frac{\$2.3 \times 10^4}{\text{acre}}$

Find: price ($/acre) = ?

Given: 1 acre = 43 560 ft² or $\frac{1\text{ acre}}{43\,560\text{ ft}^2}$, 1 lot = 80. ft x 105 ft, 1 lot = $4500 or $\frac{1\text{ lot}}{\$4500}$

Known: area = l x w = (80. ft x 105 ft) = 8.4×10^3 ft² or $\frac{1\text{ lot}}{8.4 \times 10^3\text{ ft}^2}$

Solution: price $\left(\frac{\$}{\text{acre}}\right) = \left(\frac{\$4500}{1\text{ lot}}\right)\left(\frac{1\text{ lot}}{8.4 \times 10^3\text{ ft}^2}\right)\left(\frac{43\,560\text{ ft}^2}{1\text{ acre}}\right) = \frac{\$2.3 \times 10^4}{1\text{ acre}}$

Chapter 4: Problem Solving   21

4.29 $0.0483/shave or 4.83¢/shave

Find: cost per shave ($\frac{\$}{shave}$) = ?

Given: 5 cartridges = $1.69, 1 cartridge = 7 shaves
or $\frac{5\ cartridges}{\$1.69}$, $\frac{1\ cartridge}{7\ shaves}$

Solution: cost per shave($\frac{\$}{shave}$) = ($\frac{\$1.69}{5\ cartridges}$)($\frac{1\ cartridge}{7\ shaves}$) = $\frac{\$0.0483}{1\ shave}$ or $\frac{4.83¢}{shave}$

Note: All the numbers in this problem were exact numbers. The answer could be reported to more significant figures. However, this would not be useful.

4.31 1.5×10^7 wheels

Find: # wheels = ?

Given: population of China = 1.1 billion, population Beijing= 10. million people, population of Beijing = 7.0 million cycles, 10% of cycles = tricycles or $\frac{10\ tricycles}{100\ cycles}$

Known: 1 tricycle = 3 wheels, 1 bicycle = 2 wheels, 1 million = 10^6, 1 billion = 10^9

Solution: To determine the total number of wheels, you must perform two separate calculations, one for bicycles and one for tricycles, and then sum the answers.

wheels from tricycles = ($\frac{3\ wheels}{1\ tricycle}$)($\frac{10\ tricycles}{100\ cycles}$)($7.0 \times 10^6$ cycles) = 2.1×10^6 wheels

To calculate the number of wheels from bicycles, reason that 90% of the cycles must be bicycles if 10% are tricycles or 90 bicycles = 100 cycles.

wheels from bicycles = ($\frac{2\ wheels}{1\ bicycle}$)($\frac{90\ bicycles}{100\ cycles}$)($7.0 \times 10^6$ cycles) = 1.3×10^7 wheels

total number of wheels = $1.3 \times 10^7 + 2.1 \times 10^6 = 1.5 \times 10^7$ wheels

Note: There were several unnecessary pieces of information given in this problem.

4.33 5.9×10^9 ft^3

Find: $V (ft^3) = ?$

Given: length = 3728 mi, height = 25 ft, width = 12 ft

Known: $V = l \times h \times w$, 1 mi = 5280 ft

Solution: $V(ft^3) = (25 ft)(12 ft)(\frac{5280 ft}{1 \cancel{mi}})(3728 \cancel{mi}) = 5.9 \times 10^9 ft^3$

4.35 Tokyo: $\frac{1.48 \times 10^4 \text{ people}}{\text{sq mi}}$, Japan: $\frac{898 \text{ people}}{\text{sq mi}}$

Find: pop dens$(\frac{\text{people}}{\text{sq mi}}) = ?$ (for Tokyo and for Japan)

Given: Japan: 4000 islands, 145 823 sq mi, 131 million people

Tokyo: 12.2 million people, 825 sq mi

Known: 1 million = 10^6

Solution: Tokyo: pop dens$(\frac{\text{people}}{\text{sq mi}}) = (\frac{12.2 \times 10^6 \text{ people}}{825 \text{ sq mi}}) = \frac{1.48 \times 10^4 \text{ people}}{\text{sq mi}}$

Japan: pop dens$(\frac{\text{people}}{\text{sq mi}}) = (\frac{131 \times 10^6 \text{ people}}{145\ 823 \text{ sq mi}}) = \frac{898 \text{ people}}{\text{sq mi}}$

4.37 3×10^2 lb/week

Find: weight of milk(lb)/week = ?

Given: 5 gal/da, 9 lb/gal

Known: 7 da/week

Solution: $\frac{\text{weight(lb)}}{\text{week}} = (\frac{9 \text{ lb}}{1 \cancel{gal}})(\frac{5 \cancel{gal}}{1 \cancel{da}})(\frac{7 \cancel{da}}{1 \text{ week}}) = 3 \times 10^2 \frac{\text{lb}}{\text{week}}$

4.39 40.0 m^3

Find: $V(m^3) = ?$

Given: 4.00×10^4 L, 1 m^3 = 1000 L
Solution: V(m^3) = ($\frac{1 \text{ m}^3}{1000 \text{ L}}$)(4.00 × 10^4 L) = 40.0 m^3

4.41 12 qt

Find: V (qt)/case = ?
Given: 8 bottles/case, 48 oz/bottle, 32 oz/qt
Solution: $\frac{V(qt)}{case} = (\frac{1 \text{ qt}}{32 \text{ oz}})(\frac{48 \text{ oz}}{1 \text{ bottle}})(\frac{8 \text{ bottles}}{1 \text{ case}}) = \frac{12 \text{ qt}}{case}$

4.43 1.04 ¢/page

Find: cost (¢)/page = ?
Given: 1 box = 6 ribbons, 6 ribbons = $16.60, 1 ribbon = 265 pages
Known: $1.00 = 100 ¢
Solution: $\frac{cost(¢)}{page} = (\frac{100 ¢}{\$1.00})(\frac{\$16.60}{6 \text{ ribbons}})(\frac{1 \text{ ribbon}}{265 \text{ pages}}) = \frac{1.04 ¢}{page}$

4.45 2 × 10^2 gal

Find: V (gal) = ?
Given: 1 qt water = 1 oz salt, 50 lb salt, 4 qt/gal, 16 oz/lb
Solution: V(gal) = ($\frac{1 \text{ gal}}{4 \text{ qt}}$)($\frac{1 \text{ qt}}{1 \text{ oz}}$)($\frac{16 \text{ oz}}{1 \text{ lb}}$)(50 lb) = 2 × 10^2 gal

4.47 $1.10 × 10^3/hr

Find: pay ($/hr) = ?

Given: $440 000/yr, 145 games/yr, 2 hr 45 min/game

Known: 60 min/hr

Solution: First convert the time per game into units of hours. 45 minutes = 3/4 hour = 0.75 hr. Therefore the total time per game is 2.75 hr.

$$\text{pay}\left(\frac{\$}{\text{hr}}\right) = \left(\frac{\$440\,000}{1\,\text{yr}}\right)\left(\frac{1\,\text{yr}}{145\,\text{games}}\right)\left(\frac{1\,\text{game}}{2.75\,\text{hr}}\right) = \frac{\$1.10 \times 10^3}{\text{hr}}$$

4.49 0.869 g/cm³, floats on water

Find: D (g/cm³) = ?

$$D = \frac{m}{V} = \frac{88.6\,\text{g}}{102\,\text{cm}^3} = 0.869\,\text{g/cm}^3$$

Since the density of the apple is less than the density of water, the apple would float on water.

4.51 0.354 cm³

Find: V (cm³) = ?

Given: m = 4.82 g

Known: $D = \frac{m}{V} = 13.6\,\text{g/cm}^3$

Solution: $V \times D = \frac{m}{V} \times V$

$V \times \frac{D}{D} = \frac{m}{D}$

$V = \frac{m}{D}$

$V = \frac{4.82\,\text{g}}{13.6\,\text{g/cm}^3} = \frac{4.82\,\text{g}}{13.6} \times \frac{\text{cm}^3}{\text{g}} = 0.354\,\text{cm}^3$

4.53 6.80 x 10² g

Find: mass (g) = ?

$$m = D \times V$$
$$= \left(\frac{13.6 \text{ g}}{1 \text{ mL}}\right)(50.0 \text{ mL}) = 6.80 \times 10^2 \text{ g}$$

4.55   198 g

Find: mass (g) = ?
Given: 250. mL, D = 0.791 g/mL
Known: $D = \frac{m}{V}$

$$D \times V = \frac{m}{V} \times V$$

$$D \times V = m$$

Solution: $m (g) = D \times V = \frac{0.791 \text{ g}}{1 \text{ mL}} \times 250. \text{ mL} = 198 \text{ g}$

4.57 1.1×10^2 g

Find: mass (g) = ?
Given: 21.5 g/cm³, 1 tsp, 1 tsp = 5.0 cm³
Known: m = D × V
Solution: $m (g) = \left(\frac{21.5 \text{ g}}{1 \text{ cm}^3}\right)(1 \text{ tsp})\left(\frac{5.0 \text{ cm}^3}{1 \text{ tsp}}\right) = 1.1 \times 10^2 \text{ g}$

4.59   49 g

Find: mass (g) = ?
Given: D = 0.832 g/mL, 1/4 cup, 1 qt = 4 cups, 1 L = 1.0567 qt, 1000 mL = 1 L
Known: m = D × V

Solution: First convert the volume into the unit, mL:

$$? \text{ mL} = \left(\frac{1000 \text{ mL}}{1 \text{ L}}\right)\left(\frac{1 \text{ L}}{1.0567 \text{ qt}}\right)\left(\frac{1 \text{ qt}}{4 \text{ cup}}\right)(0.25 \text{ cup}) = 59 \text{ mL}$$

Then solve the algebra problem:
$$m = D \times V = \frac{0.832 \text{ g}}{1 \text{ mL}} \times 59 \text{ mL} = 49 \text{ g}$$

Note: Based upon practical experience, you may assume two significant figures for 1/4 cup.

4.61 102.9°F

Find: $T_F = ?$ °F

$T_F = 1.8 \, T_C + 32$

$= (1.8)(39.4) + 32 = 70.9 + 32 = 102.9$°F

4.63 53°C

Find: $T_C = ?$ °C

Given: 128°F

Known: $T_F = 1.8 \, T_C + 32$. Rearranging to isolate T_C gives $T_C = \frac{(T_F - 32)}{1.8}$

Solution: $T_C = \frac{(128 - 32)}{1.8} = \frac{96}{1.8} = 53$°C

4.65 7.2×10^2 J

Find: heat (J) = ?

Known: m = 40.0 g, ΔT = 45°C - 25°C = 20.°C, SH = 0.900 J/g°C

Solution: q = (SH)(m)(ΔT) = $\left(\frac{0.900 \text{ J}}{1 \text{ g·°C}}\right)$(40.0 g)(20.°C) = 7.2×10^2 J

4.67 60. g

Find: mass (g) = ?

Given: $\Delta T = 95°C - 25°C = 70.°C$, 4.22 kcal

Known: SH(water) = 1.00 cal/g·°C, 1000 cal = 1 kcal

$q = (SH)(m)(\Delta T)$

Solution: Rearrange the algebra equation to isolate mass:

$$\frac{q}{(SH)(\Delta T)} = \frac{(SH)(m)(\Delta T)}{(SH)(\Delta T)}$$

$$\therefore m = \frac{q}{(SH)(\Delta T)} = \frac{4.22 \times 10^3 \text{ cal}}{(1.00 \text{ cal/g·°C})(70.°C)} = 60. \text{ g}$$

4.69 0.014 g or 1.4×10^{-2} g

Find: mass (g) = ?

Given: 55 mg/m^3, 0.25 m^3

Known: 1000 mg/g

Solution: $m (g) = (\frac{1 \text{ g}}{1000 \text{ mg}})(\frac{55 \text{ mg}}{1 \text{ m}^3})(0.25 \text{ m}^3) = 0.014$ g or 1.4×10^{-2} g

4.71 4.0×10^5 particles

Find: # dust particles = ?

Given: 2.5 µm/particle, 1.0 m

Known: 1 µm = 10^{-6} m

Solution: particles = $(\frac{1 \text{ particle}}{2.5 \text{ µm}})(\frac{1 \text{ µm}}{1 \times 10^{-6} \text{ m}})(1.0 \text{ m}) = 4.0 \times 10^5$ particles

4.73 465 K, 378°F

Find: $T_K = ?\ K;\ T_F = ?\ °F$

Solution: $T_K = T_C + 273 = 192 + 273 = 465\ K$

$T_F = 1.8\ T_C + 32 = (1.8)(192) + 32 = 346 + 32 = 378°F$

4.75 1.6×10^8 g

Find: mass (g) = ? g

Given: 3.0 cm, area = 62 m x 86 m, D = 0.998 g/mL

Known: D = m/V, V = area x depth, 1 cm³ = 1 mL, 100 cm = 1 m

Solution: First solve for the volume of the water in mL:

$V(mL) = (3.0\ \text{cm})(62\ \text{m} \times \frac{100\ \text{cm}}{1\ \text{m}})(86\ \text{m} \times \frac{100\ \text{cm}}{1\ \text{m}})(\frac{1\ \text{mL}}{1\ \text{cm}^3}) = 1.6 \times 10^8\ mL$

$m = D \times V = (\frac{0.998\ g}{1\ \text{mL}})(1.6 \times 10^8\ \text{mL}) = 1.6 \times 10^8\ g$

4.77 6.5×10^2 trucks

Find: # trucks = ?

Given: 12 tons coal/truck, 86 cars/train, 15 cars lumber/train, (86 - 15) cars coal/train, 110 tons coal/car

Solution: # trucks = $(\frac{1\ \text{truck}}{12\ \text{tons coal}})(\frac{110\ \text{tons coal}}{1\ \text{car}})(71\ \text{cars}) = 6.5 \times 10^2$ trucks

4.79 $335.01

Find: dollars ($) = ?

Given: 839 contestants, $0.50/contestant, 7 bears @ $7.50, 103 balloons @ $0.20, nonwinner suckers @ $2.25/gross, 12 dozen/gross, 12/dozen

Solution: There are five separate steps to reach the answer to this problem.

The total dollars paid to the proprietor must be calculated:
$$\$ = \left(\frac{\$0.50}{\text{contestant}}\right)(839 \text{ contestants}) = \$419.50$$

The dollar cost of each set of prizes must be calculated:
$$\text{first prize} = \$ = \left(\frac{\$7.50}{\text{bear}}\right)(7 \text{ bears}) = \$52.50$$
$$\text{second prize} = \$ = \left(\frac{\$0.20}{\text{balloon}}\right)(103 \text{ balloons}) = \$20.60$$

consolation prizes: $839 - 7 - 103 = 729$ suckers
$$\$ = \left(\frac{\$2.25}{1 \text{ gross}}\right)\left(\frac{1 \text{ gross}}{12 \text{ dozen}}\right)\left(\frac{1 \text{ dozen}}{12 \text{ suckers}}\right)(729 \text{ suckers}) = \$11.39$$

The profit equals the income minus the expenses = $419.50 - $52.50 - $20.60 - $11.39 = $335.01

Chapter 5:
The Structure of Atoms.
The Periodic Table

5.1 Dalton's atomic theory: (a) All matter is made up of atoms. (b) Atoms are indivisible and indestructible. (c) The atoms of an element are identical. (d) Atoms of one element can combine with atoms of another element in simple whole number ratios.

5.3 law of multiple proportions: If elements combine to form more than one compound, the ratio of the masses of one element that combine with a given mass of the other element will be a small whole number.

5.5

particle	symbol	mass, amu	charge
proton	*p*	1	+1
electron	e⁻	0	-1
neutron	n	1	0

5.7 conclusions from Rutherford's experiments: (a) Atoms contain a tiny, very dense nucleus with all the positive charge and most of the mass in the nucleus. (b) Outside the nucleus is a large region containing negative charge, equal in magnitude to the positive charge in the nucleus.

5.9 Almost all the mass is found in the *nucleus* of the atom.

5.11 The diameter of an atom is 100 000 times larger than the diameter of its nucleus.

5.13 The *atomic number* tells us the positive charge on the nucleus of an atom.

5.15 The numbers are mass numbers (sum of protons plus neutrons). Atomic number = 8.

5.17 (a) $^{4}_{2}He$ (b) $^{14}_{6}C$ (c) $^{244}_{94}Pu$ (d) $^{10}_{4}Be$ (e) $^{39}_{19}K$ (f) $^{37}_{17}Cl$

5.19

element	nuclear symbol	atomic number	mass number	Number of protons	electrons	neutrons
helium	$^{3}_{2}He$	2	3	2	2	1
sodium	$^{22}_{11}Na$	11	22	11	11	11
chlorine	$^{35}_{17}Cl$	17	35	17	17	18
copper	$^{65}_{29}Cu$	29	65	29	29	36
iron	$^{57}_{26}Fe$	26	57	26	26	31
uranium	$^{235}_{92}U$	92	235	92	92	143
aluminum	$^{27}_{13}Al$	13	27	13	13	14
fluorine	$^{19}_{9}F$	9	19	9	9	10
arsenic	$^{70}_{33}As$	33	70	33	33	37
gold	$^{197}_{79}Au$	79	197	79	79	118

5.21 The atomic mass of an element is the weighted average of the masses of the naturally occurring isotopes. It is different from the mass number of the most abundant isotope because the element is composed of several isotopes, each of which is present in a different amount and each of which contributes to the average mass of the element. Furthermore, it is possible that the most abundant isotope might contribute little more than 50% to the average mass. Thus the mass number of the most abundant isotope would not reflect the average composition of the element.

5.23   AM = 107.87 amu, silver

$$AM = (\frac{51.82}{100})(106.90 \text{ amu}) + (\frac{48.18}{100})(108.90 \text{ amu}) = 55.40 \text{ amu} + 52.47 \text{ amu} = 107.87 \text{ amu, silver.}$$

Note: Since this problem involves the combined operations of multiplication followed by addition, to determine the correct number of significant figures for the answer requires writing the results of each operation to the correct number of significant figures before beginning the next mathematical operation.

5.25   12.01 amu

$$AM = (\frac{98.89}{100})(12.000 \text{ amu}) + (\frac{1.11}{100})(13.003 \text{ amu}) = 11.87 \text{ amu} + 0.144 \text{ amu} = 12.01 \text{ amu.}$$

(See the note at the end of the solution for problem 5.23 for an explanation of the significant figures in the answer.)

5.27 2.009×10^{-23} g, mass decreases when subatomic particles are combined to make a nucleus.

Find: Mass of an atom of carbon-12 by summing the masses of its particles.
Known: One atom of carbon contains 6 protons, 6 electrons, and 6 neutrons.
Solution: AM = $(6)(1.673 \times 10^{-24} \text{ g}) + (6)(9.109 \times 10^{-28} \text{ g}) + (6)(1.675 \times 10^{-24} \text{ g}) = 2.009 \times 10^{-23}$ g

The actual mass of a carbon-12 atom is less than the calculated mass because of the mass defect. When subatomic particles are forced together to make a nucleus, some of the mass is converted to energy. This type of process taking place in stars, called fusion, is consistent with the overall conservation of mass plus energy.

5.29 5.309×10^{-23} g

    Find: mass sulfur-32 atom (g) = ?

    Given: mass carbon-12 atom = 1.9926×10^{-23} g

              mass sulfur-32 atom = 31.97 amu

    Known: mass carbon-12 atom = 12.00 amu

              12.00 amu = 1.9926×10^{-23} g   or $\dfrac{1.9926 \times 10^{-23} \text{ g}}{12.00 \text{ amu}}$

    Solution: mass (g) = $\left(\dfrac{1.9926 \times 10^{-23} \text{ g}}{12.00 \text{ amu}}\right)(31.97 \text{ amu}) = 5.309 \times 10^{-23}$ g

5.31 Mendeleev and Meyer

Dmitri Mendeleev and Lothar Meyer independently published periodic tables in which they organized the elements according to their atomic masses.

5.33 The periodic law states that similar chemical and physical properties of the elements recur periodically when the elements are arranged according to their atomic numbers.

5.35 (a) halogens in Group 7A   (b) noble gases in Group 8A   (c) alkaline earth metals in Group 2A
      (d) actinides (numbers 89 - 102) in a block at the bottom of the table in period 7

5.37 Be, Mg, Ca, Sr, Ba, Ra

The alkaline earth metals are found in Group 2A.

5.39 Cl, F, I, Br

The halogens are found in Group 7A, which is the second column from the right in the periodic table. Look for the symbols of those elements in the list.

5.41 **(a)** 7A, halogens; **(b)** 8A, noble gases; **(c)** 1A, alkali metals; **(d)** 2A, alkaline earth metals

5.43 shiny or lustrous, good heat conductor, good electrical conductor, malleable, and ductile

Physical properties can be observed without a change in the identity of a substance. All of the above properties of metals are physical properties.

5.45 tendency to gain electrons in a chemical reaction

The chemical property that distinguishes nonmetals from metals is the tendency to gain electrons.

5.47 ionic compound, e.g., sodium chloride

Since metals lose electrons to form cations and nonmetals gain electrons to form anions, they react with one another to form ionic compounds.

5.49 (a) N   (b) Br   (c) F

Nonmetals are found on the upper right side of the periodic table.

5.51 (a) Ge   (b) Sb   (c) As

Metalloids are found at the border of the line in the periodic table dividing metals from nonmetals.

5.53 Ca, Ba, Mg

The elements in Group 2A of the periodic table form +2 ions. Ca, Ba, and Mg are all alkaline earth metals from Group 2A.

5.55 (a) O^{2-} (b) N^{3-} (c) Al^{3+}

(a) Oxygen is a nonmetal. Therefore it forms an anion with a negative charge equal to 8 - group number = 8 - 2 = -2. (b) Nitrogen is also a nonmetal. It forms an anion with a charge of 8 - 5 = -3. (c) Aluminum is a metal. Metals from Groups 1A, 2A, and 3A form cations with positive charge equal to the group number, which is 3 for aluminum.

5.57 (a) $2 K + Br_2 \rightarrow 2 KBr$ (made up of K^+ and Br^-)

(b) $4 Li + O_2 \rightarrow 2 Li_2O$ (made up of Li^+ and O^{2-})

(c) $Ca + S \rightarrow CaS$ (made up of Ca^{2+} and S^{2-})

(a) Since each ion has a charge of magnitude 1, the ratio of cation to anion in the formula of the product is 1 to 1. Bromine is a diatomic element. (b) To balance the charges, you need 2 lithium ions for each oxide ion: Li_2O. Since oxygen is diatomic, you must form 2 Li_2O in order to balance the oxygen atoms. Therefore to balance lithium, you need 4 Li. (c) This was the easiest to balance because neither of the elements is diatomic and the cation and anion happen to have the same magnitude of charge. Therefore the compound formed has a 1 to 1 ratio of cation to anion, CaS.

5.59   different chemical properties: (a) and (d);   similar chemical properties: (b) and (c)

Elements in the same group have similar chemical properties. Elements in different groups have different chemical properties. The pairs of elements in (b) and (c) are in the same group. The pairs of elements in (a) and (d) are in different chemical groups.

5.61   (a) $2 Na + S \rightarrow Na_2S$ (b) $2 Ca + O_2 \rightarrow 2 CaO$ (c) $Mg + I_2 \rightarrow MgI_2$

(a) The formula for sodium sulfide requires a 2 to 1 ratio of cations to anions because the charges on the cation and the anion are not equal. There is a coefficient of 2 in front of sodium to make the atoms balance on each side of the equation. (b) Oxygen is diatomic. The formula of calcium oxide is CaO, because the ions have the same magnitude of charge. Balancing oxygen requires 2 CaO on the right. Therefore 2 Ca are needed on the left side. (c) Iodine is diatomic. Since the ions have different charges, subscripts must be used in the formula to obtain an electrically neutral compound. The ratio is 1 to 2, or MgI_2.

5.63 (a) $^{70}_{30}Zn$ (b) $^{197}_{79}Au$ (c) $^{57}_{26}Fe$

The atomic numbers of the elements are obtained from the periodic table. The atomic number is written on the lower left side of the symbol with the mass number written in the upper left corner.

5.65 (a) O^{2-}: 8p, 10e- (b) Ca^{2+}: 20 p, 18 e- (c) Br^-: 35 p, 36 e- (d) Al^{3+}: 13 p, 10 e-

The number of protons is equal to the atomic number. The number of electrons is equal to the atomic number minus the charge. (a) # p = atomic number = 8, # e- = 8 - charge = 8 - (-2) = 10. (b) # p = atomic number = 20, # e- = 20 - (+2) = 18. (c) # p = atomic number = 35, # e- = 35 - (-1) = 36. (d) # p = atomic number = 13, # e- = 13 - (+3) = 10.

5.67 Sn or Pb

Luster, malleability, and thermal conductivity are properties of metals. The only metals in Group 4A are tin and lead.

5.69 B

Brittleness and semiconductor properties are characteristic of elements that are metalloids. The only metalloid in Group 3A is boron.

Chapter 6:
Electron Structure and the Periodic Table

6.3 The difference between a continuous spectrum and a bright-line emission spectrum is that a continuous spectrum is observed when so many colors of light are being emitted at the same time that they all blend together to form a rainbow. If only a few colors of light are being emitted, they are seen as individual lines of color in a bright-line emission spectrum.

6.10 The notation for principal energy levels is $n = x$, where $x = 1, 2, 3, \ldots$

6.13 $n = 1$, 1 sublevel; $n = 2$, 2 sublevels; $n = 3$, 3 sublevels; $n = 4$, 4 sublevels

6.16 An atomic orbital is a three-dimensional region about the nucleus in which there is a high probability of locating an electron. An orbit is a circular path for an electron.

6.20 s-sublevel, 1 orbital; p-sublevel, 3 orbitals; d-sublevel, 5 orbitals; f-sublevel, 7 orbitals

6.24 two spin states possible

6.32 (a) Al: $1s^2 2s^2 2p^6 3s^2 3p^1$ (b) P: $1s^2 2s^2 2p^6 3s^2 3p^3$ (c) Cl: $1s^2 2s^2 2p^6 3s^2 3p^5$

6.34 (a) Li: $1s^22s^1$ (b) Na: $1s^22s^22p^63s^1$ (c) K: $1s^22s^22p^63s^23p^64s^1$

These elements all have one valence electron and are all found in Group 1A of the periodic table.

6.35 (a) F: $1s^22s^22p^5$ (b) Cl: $1s^22s^22p^63s^23p^5$ (c) Br: $1s^22s^22p^63s^23p^64s^23d^{10}4p^5$

The similarity is that all three elements have 5 electrons in the outermost occupied p sublevel and 7 electrons in the outermost occupied principal energy level, which is known as the valence shell. All of the elements are found in Group 7A in the periodic table.

6.39 The electron configurations of transition metals are similar in that they have from 1 to 10 electrons in the highest occupied d-subshell and no electrons in the next higher p-subshell.

6.42 (a) Cl: $3s^23p^5$ (b) As: $4s^24p^3$ (c) Si: $3s^23p^2$ (d) Rb: $5s^1$ (e) Sb: $5s^25p^3$

The valence shell is the outermost occupied electron shell. The number of electrons in the valence shell is equal to the group number.

6.43 (a) Sr: $[Kr]5s^2$ (b) Ga: $[Ar]4s^23d^{10}4p^1$ (c) Se: $[Ar]4s^23d^{10}4p^4$ (d) I: $[Kr]5s^24d^{10}5p^5$
(e) Sb: $[Kr]5s^24d^{10}5p^3$

The noble gas core is always the preceding noble gas.

6.45 (a) $4s^1$ (b) $3s^23p^5$ (c) $2s^2$ (d) $4s^24p^6$

You really need to look at the periodic table and remind yourself that the first alkali metal, alkaline earth, and halogen are found in the second period! Therefore the third alkali metal is in the fourth period. The noble gases are the only family that actually starts in the first period.

6.47 (a) Sr: (b) K• (c) :S̈: (d) :Äs•

6.49 (a) 2 unpaired electrons (d) 1 unpaired electron
(b) 1 unpaired electron (e) 0 unpaired electrons
(c) 3 unpaired electrons (f) 3 unpaired electrons

(a) :S̈e: (b) :B̈r: (c) :P̈• (d) K• (e) Ca: (f) :Äs•

6.54 (a) +1, K^+ (b) +2, Mg^{2+} (c) -1, F^- (d) -2, O^{2-} (e) -3, P^{3-} (f) +2, Ba^{2+}

For most metals in the A-Groups of the periodic table, the charge of the ion is positive and equal to the group number. For the nonmetals in the A-Groups of the periodic table, the charge of the ion is negative and equal in magnitude to (8 - group number).

6.57 (a) Ca^{2+} (b) $[:\ddot{C}l:]^-$ (c) $[:\ddot{S}:]^{2-}$ (d) K^+ (e) Al^{3+}

6.58 (a) Ca^{2+}: $1s^22s^22p^63s^23p^6$ (b) Cl^-: $1s^22s^22p^63s^23p^6$ (c) S^{2-}: $1s^22s^22p^63s^23p^6$
(d) K^+: $1s^22s^22p^63s^23p^6$ (e) Al^{3+}: $1s^22s^22p^6$

6.61 (a) Ne (b) Ne (c) Ar (d) Ar (e) Ar

Each pair of ions is isoelectronic with the noble gas that has the same number of electrons as the ion. This is the noble gas closest in atomic number to the atoms from which the ions are formed.

6.64 Rb^+, Sr^{2+}, Y^{3+}

Cations isoelectronic with a Br^- ion would come from metals having 1, 2, or 3 electrons more than the noble gas krypton.

6.66 Li• $\longrightarrow$ Li^+ + e^-

6.68 :Cl̈• + e^- $\longrightarrow$ [:C̈l:]$^-$ or $Cl_2(g) + 2e^- \rightarrow 2Cl^-$

6.69 (a)   Li•   +   :C̈l• $\longrightarrow$ Li^+ + [:C̈l:]$^-$ or LiCl

    oxidized   reduced
    reducing agent   oxidizing agent

(b)   Ca: +   :S̈• $\longrightarrow$ Ca^{2+} + [:S̈:]$^{2-}$ or CaS

    oxidized   reduced
    reducing agent   oxidizing agent

6.74 (a) Cs > K (b) I > Cl (c) Ba > Mg (d) K > Br (e) Be > O

Atomic size increases down a family, (a), (b), and (c) and decreases across a period, (d) and (e).

6.76   (a) $Cl^- > Cl$ (b) $Cl^- > Li^+$ (c) $S^{2-} > O^{2-}$

(a) Anions are always larger than the parent atom. (b) Anions are larger than their parent, and cations are smaller than their parent. Furthermore, chlorine is further down in the periodic table than lithium. The lithium ion has only one shell of electrons. The chloride ion has three shells of electrons. For all these reasons, the chloride ion is larger than the lithium ion. (c) Both sulfur and oxygen are in Group 6A. The anions will have the same size relationship as the parent atoms, and size increases going down a group in the periodic table.

6.81   (a) Na   (b) I   (c) Cl   (d) Mg

(a) Sodium and cesium are in the same group. Ionization energy is larger at the top of a group. (b-d) Ionization energy increases from left to right in the periodic table.

6.86   (a) F   (b) Cl   (c) O   (d) O   (e) Cl

Nonmetals have larger electron affinities than either metals or noble gases.

6.87   (a) F   (b) Cl   (c) O   (d) S

(a) In general, electron affinity increases from bottom to top in the periodic table. (b-d) In general, electron affinity increases from left to right in the periodic table.

6.89 (a) 7 valence electrons (b) 5 (c) 3 (d) 6 (e) 5 (f) 7

The number of valence electrons is equal to the number of s and p electrons in the highest energy level. This number is the same as the group number for A-group elements.

6.91 (a) K (b) Mg (c) Al (d) Ca

Ionization energy increases from left to right in the periodic table. Therefore it is easier for elements on the left side of the periodic table to lose electrons.

6.94 (a) Sn: $1s^22s^22p^63s^23p^64s^23d^{10}4p^65s^24d^{10}5p^2$ (b) Rb$^+$: $1s^22s^22p^63s^23p^64s^23d^{10}4p^6$
(c) Se^{2-}: $1s^22s^22p^63s^23p^64s^23d^{10}4p^6$

6.96 (a) Mg (b) As (c) Se

(a) two 3s electrons indicate that this is the element in period 3, Group 2A

(b) three 4p electrons indicate that is is the element in period 4, Group 5A

(c) four 4p electrons (two paired and two unpaired) indicate that this is the element in period 4, Group 6A

6.99 (a) O (b) O (c) O* (d) Po (e) O

The answers above are based upon general trends in the periodic table. Electron affinities don't follow the trends as well as the other properties. Thus, even though the trend in electron affinities predicts that oxygen should have a larger electron affinity, Figure 6.17 shows that sulfur actually has a larger value for its electron affinity. In the absence of a table listing specific values, you should use the general trends to make predictions.

Chapter 7
Composition and Formulas of Compounds

7.1 A mole of a substance is the amount of substance that contains the same number of formula units as there are carbon atoms in exactly 12 grams of carbon-12.

7.3 A formula unit is the atom or group of atoms represented by the formula of a substance.

The term formula unit is a generic unit that may to used to talk about any substance, without having to specify whether the units are atoms, molecules, or ions.

7.5 The molar mass is the mass in grams of one mole of any substance.

Molar mass is often abbreviated MM. It has the units g/mol.

7.7 (a) 63.6 g/mol   (c) 39.1 g/mol   (e) 40.0 g/mol
(b) 32.1 g/mol   (d) 74.9 g/mol   (f) 24.3 g/mol

The molar mass of an element is the atomic mass found in the periodic table expressed in the unit g/mol.

47

7.9 (a) 358 g/mol (b) 138 g/mol (c) 74.5 g/mol (d) 80.1 g/mol (e) 58.3 g/mol

(a) MM-$Fe_3(PO_4)_2$ = 3 (55.9 g/mol Fe) + 2 (31.0 g/mol P) + 8 (16.0 g/mol O) = 358 g/mol $Fe_3(PO_4)_2$

(b) MM-PCl_3 = 31.0 g/mol P + 3 (35.5 g/mol Cl) = 138 g/mol PCl_3

(c) MM-NaClO = 23.0 g/mol Na + 35.5 g/mol Cl + 16.0 g/mol O = 74.5 g/mol NaClO

(d) MM-SO_3 = 32.1 g/mol S + 3 (16.0 g/mol O) = 80.1 g/mol SO_3

(e) MM-$Mg(OH)_2$ = 24.3 g/mol Mg + 2 (16.0 g/mol O) + 2 (1.01 g/mol H) = 58.3 g/mol $Mg(OH)_2$

7.11 (a) 58.1 g/mol (b) 46.1 g/mol (c) 98.0 g/mol (d) 123.1 g/mol (e) 284 g/mol
(f) 44.0 g/mol (g) 58.1 g/mol

(a) MM-C_4H_{10} = 4 (12.0 g/mol C) + 10 (1.01 g/mol H) = 58.1 g/mol C_4H_{10}

(b) MM-C_2H_6O = 2 (12.0 g/mol C) + 6 (1.01 g/mol H) + 16.0 g/mol O = 46.1 g/mol C_2H_6O

(c) MM-H_3PO_4 = 3 (1.01 g/mol H) + 31.0 g/mol P + 4 (16.0 g/mol O) = 98.0 g/mol H_3PO_4

(d) MM-$C_6H_5NO_2$ = 6 (12.0 g/mol C) + 5 (1.01 g/mol H) + 14.0 g/mol N + 2 (16.0 g/mol O) = 123.1 g/mol $C_6H_5NO_2$

(e) MM-P_4O_{10} = 4 (31.0 g/mol P) + 10 (16.0 g/mol O) = 284 g/mol P_4O_{10}

(f) MM-N_2O = 2 (14.0 g/mol N) + 16.0 g/mol O = 44.0 g/mol N_2O

(g) MM-C_3H_6O = 3 (12.0 g/mol C) + 6 (1.01 g/mol H) + 16.0 g/mol O = 58.1 g/mol C_3H_6O

7.13 (a) 2.86 mol (b) 0.646 mol (c) 0.1024 mol (d) 1.385 mol

(a) Find: number of moles C_3H_8 = ?

Given: 126 g C_3H_8

Known: MM-C_3H_8 = 44.1 g/mol C_3H_8

Solution: moles C_3H_8 = ($\frac{1 \text{ mol } C_3H_8}{44.1 \text{ g } C_3H_8}$)(126 g C_3H_8) = 2.86 mol C_3H_8

(b) Find: number of moles Al_2O_3 = ?

Given: 65.9 g Al_2O_3

Known: MM-Al_2O_3 = 102 g/mol Al_2O_3

Solution: moles Al_2O_3 = ($\frac{1 \text{ mol } Al_2O_3}{102 \text{ g } Al_2O_3}$)(65.9 g Al_2O_3) = 0.646 mol Al_2O_3

(c) Find: number of moles $KClO_3$ = ?

Given: 12.56 g $KClO_3$

Known: MM-$KClO_3$ = 122.6 g/mol $KClO_3$

Solution: moles $KClO_3$ = ($\frac{1 \text{ mol } KClO_3}{122.6 \text{ g } KClO_3}$)(12.56 g $KClO_3$) = 0.1024 mol $KClO_3$

(d) Find: number of moles NH_4Br = ?

Given: 135.7 g NH_4Br

Known: MM-NH_4Br = 97.95 g/mol NH_4Br

Solution: moles NH_4Br = ($\frac{1 \text{ mol } NH_4Br}{97.95 \text{ g } NH_4Br}$)(135.7 g NH_4Br) = 1.385 mol NH_4Br

7.15 (a) 225 g (b) 353 g (c) 61.7 g (d) 80.0 g

(a) Find: mass (g) = ?

Given: 4.33 mol Cr

Known: 52.0 g/mol Cr

Solution: mass (g) = ($\frac{52.0 \text{ g Cr}}{1 \text{ mol Cr}}$)(4.33 mol Cr) = 225 g Cr

(b) Find: mass (g) = ?

Given: 8.03 mol CO_2

Known: 44.0 g/mol CO_2

Solution: mass (g) = $(\frac{44.0 \text{ g CO}_2}{1 \text{ mol CO}_2})(8.03 \text{ mol CO}_2)$ = 353 g CO_2

(c) Find: mass (g) = ?

Given: 2.54 mol Mg

Known: 24.3 g/mol Mg

Solution: mass (g) = $(\frac{24.3 \text{ g Mg}}{1 \text{ mol Mg}})(2.54 \text{ mol Mg})$ = 61.7 g Mg

(d) Find: mass (g) = ?

Given: 0.234 mol $Al_2(SO_4)_3$

Known: 342 g/mol $Al_2(SO_4)_3$

Solution: mass (g) = $(\frac{342 \text{ g Al}_2(SO_4)_3}{1 \text{ mol Al}_2(SO_4)_3})(0.234 \text{ mol Al}_2(SO_4)_3)$ = 80.0 g $Al_2(SO_4)_3$

7.17 (a) 206 g (b) 0.155 g (c) 21.7 g (d) 23.1 g

(a) Find: mass (g) = ?

Given: 1.06 mol K_2CrO_4

Known: 194 g/mol K_2CrO_4

Solution: mass (g) = $(\frac{194 \text{ g K}_2\text{CrO}_4}{1 \text{ mol K}_2\text{CrO}_4})(1.06 \text{ mol K}_2\text{CrO}_4)$ = 206 g K_2CrO_4

(b) Find: mass (g) = ?

Given: 1.50 mmol NaBr

Known: 103 g/mol NaBr, 1000 mmol/1 mol NaBr

Solution: mass (g) = $(\frac{103 \text{ g NaBr}}{1 \text{ mol NaBr}})(\frac{1 \text{ mol NaBr}}{1000 \text{ mmol NaBr}})(1.50 \text{ mmol NaBr})$ = 0.155 g NaBr

(c) Find: mass (g) = ?

Given: 0.542 mol NaOH

Known: 40.0 g/mol NaOH

Solution: mass (g) = $(\frac{40.0 \text{ g NaOH}}{1 \text{ mol NaOH}})(0.542 \text{ mol NaOH})$ = 21.7 g NaOH

Chapter 7: Composition and Formulas of Compounds 51

(d) Find: mass (g) = ?

Given: 0.385 mol CH_4N_2O

Known: 60.0 g/mol CH_4N_2O

Solution: mass (g) = $(\dfrac{60.0 \text{ g } CH_4N_2O}{1 \text{ mol } CH_4N_2O})(0.385 \text{ mol } CH_4N_2O)$ = 23.1 g CH_4N_2O

7.19 (a) 1.41×10^{22} molecules (b) 3.0×10^{21} molecules (c) 3.9×10^{23} molecules

(d) 7.53×10^{23} molecules

(a) Find: number of molecules = ?

Given: 0.0235 mol N_2

Known: 6.02×10^{23} molecules/mol

Solution: molecules = $(\dfrac{6.02 \times 10^{23} \text{ molecules } N_2}{1 \text{ mol } N_2})(0.0235 \text{ mol } N_2)$

= 1.41×10^{22} molecules N_2

(b) Find: number of molecules = ?

Given: 0.0050 mol O_3

Known: 6.0×10^{23} molecules/mol

Solution: molecules = $(\dfrac{6.0 \times 10^{23} \text{ molecules } O_3}{1 \text{ mol } O_3})(0.0050 \text{ mol } O_3)$ = 3.0×10^{21} molecules O_3

(c) Find: number of molecules = ?

Given: 0.65 mol $C_5H_{10}O_4$

Known: 6.0×10^{23} molecules/mol

Solution: molecules = $(\dfrac{6.0 \times 10^{23} \text{ molecules } C_5H_{10}O_4}{1 \text{ mol } C_5H_{10}O_4})(0.65 \text{ mol } C_5H_{10}O_4)$

= 3.9×10^{23} molecules $C_5H_{10}O_4$

(d) Find: number of molecules = ?

Given: 1.25 mol NO_2

Known: 6.02×10^{23} molecules/mol

Solution: molecules = $(\dfrac{6.02 \times 10^{23} \text{ molecules NO}_2}{1 \text{ mol NO}_2})(1.25 \text{ mol NO}_2)$

= 7.53×10^{23} molecules NO$_2$

7.21 (a) 1.00×10^{25} molecules (b) 2.33×10^{22} molecules (c) 1.2×10^{19} molecules

(a) Find: number of molecules = ?

Given: 1.00 kg HC$_2$H$_3$O$_2$

Known: 6.02×10^{23} molecules/mol, 1 kg/1000 g, 60.0 g/mol HC$_2$H$_3$O$_2$

Solution: molecules = $(\dfrac{6.02 \times 10^{23} \text{ molecules}}{1 \text{ mol}})(\dfrac{1 \text{ mol}}{60.0 \text{ g}})(\dfrac{1000 \text{ g}}{1 \text{ kg}})(1.00 \text{ kg})$

= 1.00×10^{25} molecules

(b) Find: number of molecules = ?

Given: 6.20 g Br$_2$

Known: 6.02×10^{23} molecules/mol, 160. g/mol Br$_2$

Solution: molecules = $(\dfrac{6.02 \times 10^{23} \text{ molecules Br}_2}{1 \text{ mol Br}_2})(\dfrac{1 \text{ mol Br}_2}{160. \text{ g Br}_2})(6.20 \text{ g Br}_2)$

= 2.33×10^{22} molecules Br$_2$

(c) Find: number of molecules = ?

Given: 2.9 mg C$_8$H$_8$O$_3$

Known: 6.0×10^{23} molecules/mol, 1 g/1000 mg, 150 g/mol C$_8$H$_8$O$_3$

Solution: molecules = $(\dfrac{6.0 \times 10^{23} \text{ molecules}}{1 \text{ mol}})(\dfrac{1 \text{ mol}}{150 \text{ g}})(\dfrac{1 \text{ g}}{1000 \text{ mg}})(2.9 \text{ mg})$

= 1.2×10^{19} molecules C$_8$H$_8$O$_3$

7.23 (a) 9.93×10^{19} atoms (b) 3.8×10^{16} atoms (c) 1.7×10^{19} atoms (d) 4.88×10^{25} atoms

(a) Find: number of atoms = ?

Given: 1.65×10^{-4} mol Mg

Known: 6.02×10^{23} atoms/mol

Solution: atoms = $(\dfrac{6.02 \times 10^{23} \text{ atoms}}{1 \text{ mol}})(1.65 \times 10^{-4} \text{ mol Mg}) = 9.93 \times 10^{19}$ atoms Mg

(b) Find: number of atoms O = ?

Given: 1.0 µg O_3

Known: 1×10^6 µg/1 g, 48 g/mol O_3, 6.0×10^{23} molecules/mol O_3,

3 atoms O/1 molecule O_3

Solution: atoms O = $(\dfrac{3 \text{ atoms O}}{1 \text{ molecule } O_3})(\dfrac{6.0 \times 10^{23} \text{ molecules}}{1 \text{ mol}})(\dfrac{1 \text{ mol}}{48 \text{ g}})(\dfrac{1 \text{ g}}{1 \times 10^6 \text{ µg}})(1 \text{ µg})$

= 3.8×10^{16} atoms

(c) Find: number of atoms = ?

Given: 3.2 mg Ag

Known: 6.0×10^{23} atoms/mol, 110 g/mol Ag, 1000 mg/g

Solution: atoms = $(\dfrac{6.0 \times 10^{23} \text{ atoms Ag}}{1 \text{ mol Ag}})(\dfrac{1 \text{ mol Ag}}{110 \text{ g Ag}})(\dfrac{1 \text{ g Ag}}{1000 \text{ mg Ag}})(3.2 \text{ mg Ag})$

= 1.7×10^{19} atoms Ag

(d) Find: number of atoms = ?

Given: 2.60 kg S

Known: 6.02×10^{23} atoms/mol, 32.1 g/mol S, 1000 g/kg

Solution: atoms = $(\dfrac{6.02 \times 10^{23} \text{ atoms S}}{1 \text{ mol S}})(\dfrac{1 \text{ mol S}}{32.1 \text{ g S}})(\dfrac{1000 \text{ g S}}{1 \text{ kg S}})(2.60 \text{ kg S})$

= 4.88×10^{25} atoms S

7.25 (a) 5.66×10^{-23} g   (b) 7.31×10^{-23} g   (c) 7.33×10^{-23} g   (d) 2.23×10^{-22} g

(a) Find: mass (g) = ?

Given: 1 molecule H_2S (exactly)

Known: MM-H₂S = 34.1 g/mol

Solution: mass (g) = $(\frac{34.1 \text{ g}}{1 \text{ mol}})(\frac{1 \text{ mol}}{6.02 \times 10^{23} \text{ molecules}})(1 \text{ molecule}) = 5.66 \times 10^{-23}$ g

(with 6.02 x 10²³ molecules/mol)

(b) Find: mass (g) = ?

Given: 1 molecule CO₂ (exactly)

Known: MM-CO₂ = 44.0 g/mol

Solution: mass (g) = $(\frac{44.0 \text{ g}}{1 \text{ mol}})(\frac{1 \text{ mol}}{6.02 \times 10^{23} \text{ molecules}})(1 \text{ molecule}) = 7.31 \times 10^{-23}$ g

(with 6.02 x 10²³ molecules/mol)

(c) Find: mass (g) = ?

Given: 1 molecule C₃H₈ (exactly)

Known: MM-C₃H₈ = 44.1 g/mol

Solution: mass (g) = $(\frac{44.1 \text{ g}}{1 \text{ mol}})(\frac{1 \text{ mol}}{6.02 \times 10^{23} \text{ molecules}})(1 \text{ molecule}) = 7.33 \times 10^{-23}$ g

(with 6.02 x 10²³ molecules/mol)

(d) Find: mass (g) = ?

Given: 1 molecule C₂H₃Cl₃ (exactly)

Known: MM-C₂H₃Cl₃ = 134 g/mol

Solution: mass (g) = $(\frac{134 \text{ g}}{1 \text{ mol}})(\frac{1 \text{ mol}}{6.02 \times 10^{23} \text{ molecules}})(1 \text{ molecule}) = 2.23 \times 10^{-22}$ g

(with 6.02 x 10²³ molecules/mol)

7.27 (a) 6.66 x 10⁻²³ g (b) 9.75 x 10⁻²³ g (c) 1.98 x 10⁻²² g (d) 1.15 x 10⁻²³ g
(e) 4.67 x 10⁻²³ g (f) 1.24 x 10⁻²² g

(a) Find: mass (g) = ?

Given: 1 atom Ca (exactly)

Known: MM-Ca = 40.1 g/mol

Chapter 7: Composition and Formulas of Compounds   55

Solution: mass (g) = $(\frac{40.1 \text{ g}}{1 \text{ mol}})(\frac{1 \text{ mol}}{6.02 \times 10^{23} \text{ atoms}})$(1 atom) = 6.66 × 10⁻²³ g

(b) Find: mass (g) = ?

Given: 1 atom Ni (exactly)

Known: MM-Ni = 58.7 g/mol

Solution: mass (g) = $(\frac{58.7 \text{ g}}{1 \text{ mol}})(\frac{1 \text{ mol}}{6.02 \times 10^{23} \text{ atoms}})$(1 atom) = 9.75 × 10⁻²³ g

(c) Find: mass (g) = ?

Given: 1 atom Sn (exactly)

Known: MM-Sn = 119 g/mol

Solution: mass (g) = $(\frac{119 \text{ g}}{1 \text{ mol}})(\frac{1 \text{ mol}}{6.02 \times 10^{23} \text{ atoms}})$(1 atom) = 1.98 × 10⁻²² g

(d) Find: mass (g) = ?

Given: 1 atom Li (exactly)

Known: MM-Li = 6.94 g/mol

Solution: mass (g) = $(\frac{6.94 \text{ g}}{1 \text{ mol}})(\frac{1 \text{ mol}}{6.02 \times 10^{23} \text{ atoms}})$(1 atom) = 1.15 × 10⁻²³ g

(e) Find: mass (g) = ?

Given: 1 atom Si (exactly)

Known: MM-Si = 28.1 g/mol

Solution: mass (g) = $(\frac{28.1 \text{ g}}{1 \text{ mol}})(\frac{1 \text{ mol}}{6.02 \times 10^{23} \text{ atoms}})$(1 atom) = 4.67 × 10⁻²³ g

(f) Find: mass (g) = ?

Given: 1 atom As (exactly)

Known: MM-As = 74.9 g/mol

6.02 x 10^{23} atoms/mol

Solution: mass (g) = ($\frac{74.9 \text{ g}}{1 \text{ mol}}$)($\frac{1 \text{ mol}}{6.02 \times 10^{23} \text{ atoms}}$)(1 atom) = 1.24 x 10^{-22} g

7.29 4.2 x 10^{-18} mol

Find: mol = ?

Given: 2.5 x 10^6 atoms

Known: 6.0 x 10^{23} atoms/mol

Solution: mol = ($\frac{1 \text{ mol}}{6.0 \times 10^{23} \text{ atoms}}$)(2.5 x 10^6 atoms) = 4.2 x 10^{-18} mol

7.31 (a) 39.3% Na, 60.7% Cl (b) 7.79% C, 92.2% Cl (c) 23.6% K, 76.5% I

(d) 27.4% Na, 1.20% H, 14.3% C, 57.1% O

(a) % Na = ($\frac{23.0 \text{ g Na}}{58.5 \text{ g NaCl}}$)(100) = 39.3%; % Cl = ($\frac{35.5 \text{ g}}{58.5 \text{ g}}$)(100) = 60.7%

(b) % C = ($\frac{12.0 \text{ g}}{154 \text{ gCCl}_4}$)(100) = 7.79%; % Cl = ($\frac{(4)(35.5 \text{ g})}{154 \text{ g}}$)(100) = 92.2%

(c) % K = ($\frac{39.1 \text{ g}}{166 \text{ g KI}}$)(100) = 23.6%; % I = ($\frac{127 \text{ g}}{166 \text{ g}}$)(100) = 76.5%

(d) %Na = ($\frac{23.0 \text{ g}}{84.0 \text{ g NaHCO}_3}$)(100) = 27.4%; % H = ($\frac{1.01 \text{ g}}{84.0 \text{ g}}$)(100) = 1.20%;

% C = ($\frac{12.0 \text{ g}}{84.0 \text{ g}}$)(100) = 14.3 %, % O = ($\frac{(3)(16.0 \text{ g})}{84.0 \text{ g}}$)(100) = 57.1 %

7.33 (a) 81.6% (b) 47.7% (c) 83.5% (d) 50.0% (e) 44.1%

(a) % Cl = ($\frac{(2)(35.5 \text{ g Cl})}{87.0 \text{ g Cl}_2\text{O}}$)(100) = 81.6%

(b) % Cl = ($\frac{35.5 \text{ g Cl}}{74.5 \text{ g NaClO}}$)(100) = 47.7%

(c) $\% \text{ Cl} = (\dfrac{(2)(35.5 \text{ g Cl})}{85.0 \text{ g CH}_2\text{Cl}_2})(100) = 83.5\%$

(d) $\% \text{ Cl} = (\dfrac{(5)(35.5 \text{ g Cl})}{355 \text{ g C}_{14}\text{H}_9\text{Cl}_5})(100) = 50.0\%$

(e) $\% \text{ Cl} = (\dfrac{(4)(35.5 \text{ g Cl})}{322 \text{ g C}_{12}\text{H}_4\text{Cl}_4\text{O}_2})(100) = 44.1\%$

7.35 46.7% N

$\% \text{ N} = (\dfrac{(2)(14.0 \text{ g N})}{60.0 \text{ g CH}_4\text{N}_2\text{O}})(100) = 46.7\%$

7.37 (a) 93.1% Ag (b) 63.5% (c) 24.7% K (d) 28.5% Cu (e) 38.0% Co (f) 68.4% Cr

(a) $\% \text{ Ag} = (\dfrac{(2)(108 \text{ g Ag})}{232 \text{ g Ag}_2\text{O}})(100) = 93.1\%$

(b) $\% \text{ Fe} = (\dfrac{55.9 \text{ g Fe}}{88.0 \text{ g FeS}})(100) = 63.5\%$

(c) $\% \text{ K} = (\dfrac{39.1 \text{ g K}}{158.0 \text{ g KMnO}_4})(100) = 24.7\%$

(d) $\% \text{ Cu} = (\dfrac{63.6 \text{ g Cu}}{223 \text{ g CuBr}_2})(100) = 28.5\%$

(e) $\% \text{ Co} = (\dfrac{58.9 \text{ g Co}}{155 \text{ g CoSO}_4})(100) = 38.0\%$

(f) $\% \text{ Cr} = (\dfrac{(2)(52.0 \text{ g Cr})}{152 \text{ g Cr}_2\text{O}_3})(100) = 68.4\%$

7.39 (d), (e), and (f) are empirical formulas.

(a) Na_2O_2 can be reduced to NaO; therefore it is not an empirical formula.

(b) $HC_2H_3O_2$ can be rewritten as $C_2H_4O_2$, which can be reduced to CH_2O.

(c) C_2H_2 can be reduced to CH.

7.41 (b) HO (d) CH$_2$O

(a), (c), (e), (f), and (g) can not be further reduced. Therefore, they are already empirical formulas.

7.43 Fe$_2$S$_3$

Element	Step 1: mass, g	Step 2: number of moles	Step 3: simplest whole-number mole ratio	Step 4: formula
iron	3.65 g	$\frac{3.65 \text{ g}}{55.9 \text{ g/mol}} = 0.0653$ mol	$\frac{0.0653 \text{ mol}}{0.0653 \text{ mol}} = 1.00$ or 2	Fe$_2$S$_3$
sulfur	3.15 g	$\frac{3.15 \text{ g}}{32.1 \text{ g/mol}} = 0.0981$ mol	$\frac{0.0981 \text{ mol}}{0.0653 \text{ mol}} = 1.50$ or 3	

7.45 NaMnO$_4$

Element	Step 1: mass, g	Step 2: number of moles	Step 3: simplest whole-number mole ratio	Step 4: formula
sodium	0.0412 g	$\frac{0.0412 \text{ g}}{23.0 \text{ g/mol}} = 0.00179$ mol	$\frac{0.00179 \text{ mol}}{0.00179 \text{ mol}} = 1.00$ or 1	NaMnO$_4$
manganese	0.0980 g	$\frac{0.0980 \text{ g}}{54.9 \text{ g/mol}} = 0.00179$ mol	$\frac{0.00170 \text{ mol}}{0.00179 \text{ mol}} = 1.00$ or 1	
oxygen	0.115 g*	$\frac{0.115 \text{ g}}{16.0 \text{ g/mol}} = 0.00719$ mol	$\frac{0.00719 \text{ mol}}{0.00179 \text{ mol}} = 4.02$ or 4	

* mass of oxygen = mass of compound - (mass of Na + Mn) = 0.254 g - (0.0412 g + 0.0980 g)

= 0.115 g

7.47 $C_7H_7NO_2$

Element	Step 1: mass, g	Step 2: number of moles	Step 3: simplest whole-number mole ratio	Step 4: formula
carbon	61.3 g	$\frac{61.3 \text{ g}}{12.0 \text{ g/mol}} = 5.11$ mol	$\frac{5.11 \text{ mol}}{0.729 \text{ mol}} = 7.01$ or 7	$C_7H_7NO_2$
hydrogen	5.11 g	$\frac{5.11 \text{ g}}{1.01 \text{ g/mol}} = 5.06$ mol	$\frac{5.06 \text{ mol}}{0.729 \text{ mol}} = 6.94$ or 7	
oxygen	23.4 g	$\frac{23.4 \text{ g}}{16.0 \text{ g/mol}} = 1.46$ mol	$\frac{1.46 \text{ mol}}{0.729 \text{ mol}} = 2.00$ or 2	
nitrogen	10.2 g	$\frac{10.2 \text{ g}}{14.0 \text{ g/mol}} = 0.729$ mol	$\frac{0.729 \text{ mol}}{0.729 \text{ mol}} = 1.00$ or 1	

7.49 NO_2

Element	Step 1: mass, g	Step 2: number of moles	Step 3: simplest whole-number mole ratio	Step 4: formula
nitrogen	30.4 g	$\frac{30.4 \text{ g}}{14.0 \text{ g/mol}} = 2.17$ mol	$\frac{2.17 \text{ mol}}{2.17 \text{ mol}} = 1.00$ or 1	NO_2
oxygen	69.6 g *	$\frac{69.6 \text{ g}}{16.0 \text{ g/mol}} = 4.35$ mol	$\frac{4.35 \text{ mol}}{2.17 \text{ mol}} = 2.00$ or 2	

* mass of oxygen: 100% - 30.4% N = 69.6% O

7.51 N_2O_3

Element	Step 1: mass, g	Step 2: number of moles	Step 3: simplest whole-number mole ratio	Step 4: formula
nitrogen	36.8 g	$\frac{36.8 \text{ g}}{14.0 \text{ g/mol}} = 2.63$ mol	$\frac{2.63 \text{ mol}}{2.63 \text{ mol}} = 1.00 \times 2 = 2$	N_2O_3
oxygen	63.1 g	$\frac{63.1 \text{ g}}{16.0 \text{ g/mol}} = 3.94$ mol	$\frac{3.94 \text{ mol}}{2.63 \text{ mol}} = 1.50 \times 2 = 3$	

7.53 FeS_2

Element	Step 1: mass, g	Step 2: number of moles	Step 3: simplest whole-number mole ratio	Step 4: formula
iron	0.730 g	$\frac{0.730 \text{ g}}{55.9 \text{ g/mol}} = 0.0131 \text{ mol}$	$\frac{0.0131 \text{ mol}}{0.0131 \text{ mol}} = 1.00 \text{ or } 1$	FeS_2
sulfur	0.840 g	$\frac{0.840 \text{ g}}{32.1 \text{ g/mol}} = 0.0262 \text{ mol}$	$\frac{0.0262 \text{ mol}}{0.0131 \text{ mol}} = 2.00 \text{ or } 2$	

7.55 CH_3

Element	Step 1: mass, g	Step 2: number of moles	Step 3: simplest whole-number mole ratio	Step 4: formula
carbon	not needed	0.0334 mol* (see below)	$\frac{0.0334 \text{ mol}}{0.0334 \text{ mol}} = 1.00 \text{ or } 1$	CH_3
hydrogen	not needed	0.0990 mol* (see below)	$\frac{0.0990 \text{ mol}}{0.0334 \text{ mol}} = 2.96 \text{ or } 3$	

Find: moles C = ?; moles H = ?

Given: 1.47 g CO_2; 0.891 g H_2O

Known: 1 mol C per mol CO_2; 2 mol H per mol H_2O

44.0 g/mol CO_2; 18.0 g/mol H_2O

Solution:
moles C = $(\frac{1 \text{ mol C}}{1 \text{ mol } CO_2})(\frac{1 \text{ mol } CO_2}{44.0 \text{ g } CO_2})(1.47 \text{ g } CO_2) = 0.0334$ mol C

moles H = $(\frac{2 \text{ mol H}}{1 \text{ mol } H_2O})(\frac{1 \text{ mol } H_2O}{18.0 \text{ g } H_2O})(0.891 \text{ g } H_2O) = 0.0990$ mol H

7.57 empirical formula and molar mass

7.59 $C_6H_8O_7$

Find: molecular formula

Given: MM = 192 g/mol

37.5% C, 4.20% H, 58.3% O

Known: $x = (\frac{MM}{EFM})$, where EFM is the empirical formula mass for citric acid

Solution: First calculate the empirical formula for citric acid from the percentage composition:

Element	Step 1: mass, g	Step 2: number of moles	Step 3: simplest whole-number mole ratio	Step 4: formula
carbon	37.5 g	$\frac{37.5 \text{ g}}{12.0 \text{ g/mol}} = 3.13$ mol	$\frac{3.13 \text{ mol}}{3.13 \text{ mol}} = 1.00 \times 6 = 6$	$C_6H_8O_7$
hydrogen	4.20 g	$\frac{4.20 \text{ g}}{1.01 \text{ g/mol}} = 4.16$ mol	$\frac{4.16 \text{ mol}}{3.13 \text{ mol}} = 1.33 \times 6 = 8$	
oxygen	58.3 g	$\frac{58.3 \text{ g}}{16.0 \text{ g/mol}} = 3.64$ mol	$\frac{3.64 \text{ mol}}{3.13 \text{ mol}} = 1.16 \times 6 = 7$	

For $C_6H_8O_7$, EFM = 192 g/mol.

To calculate x, the number of formula units per molecule:
$$x = (\frac{MM}{EFM}) = (\frac{192}{192}) = 1.00$$

The molecular formula is $C_6H_8O_7$.

7.61 $C_4H_{12}N_2$

Find: molecular formula

Given: MM = 88.0 g/mol

54.5% C, 13.8% H, 31.8% N

Known: $x = (\frac{MM}{EFM})$, where EFM is the empirical formula mass for putrescine

Solution: First calculate the empirical formula for putrescine from the percentage composition:

Element	Step 1: mass, g	Step 2: number of moles	Step 3: simplest whole-number mole ratio	Step 4: formula
carbon	54.5 g	$\frac{54.5 \text{ g}}{12.0 \text{ g/mol}} = 4.54$ mol	$\frac{4.54 \text{ mol}}{2.27 \text{ mol}} = 2.00$ or 2	C_2H_6N
hydrogen	13.8 g	$\frac{13.8 \text{ g}}{1.01 \text{ g/mol}} = 13.7$ mol	$\frac{13.7 \text{ mol}}{2.27 \text{ mol}} = 6.04$ or 6	
nitrogen	31.8 g	$\frac{31.8 \text{ g}}{14.0 \text{ g/mol}} = 2.27$ mol	$\frac{2.27 \text{ mol}}{2.27 \text{ mol}} = 1.00$ or 1	

For C_2H_6N, EFM = 44.1 g/mol.

To calculate x, the number of formula units per molecule:
$$x = \left(\frac{MM}{EFM}\right) = \left(\frac{88.0}{44.1}\right) = 2.00 \text{ or } 2$$

The molecular formula is $(C_2H_6N)_2$ or $C_4H_{12}N_2$.

7.63 $C_{10}H_8$

Find: molecular formula

Given: MM = 128 g/mol

Known: 93.7% C, 6.29% H
$x = \left(\frac{MM}{EFM}\right)$, where EFM is the empirical formula mass for naphthalene

Solution: First calculate the empirical formula for naphthalene from the percentage composition:

Element	Step 1: mass, g	Step 2: number of moles	Step 3: simplest whole-number mole ratio	Step 4: formula
carbon	93.7 g	$\frac{93.7 \text{ g}}{12.0 \text{ g/mol}} = 7.81$ mol	$\frac{7.81 \text{ mol}}{6.23 \text{ mol}} = 1.25 \times 4 = 5$	C_5H_4

| hydrogen | 6.29 g | $\frac{6.29 \text{ g}}{1.01 \text{ g/mol}}$ = 6.23 mol | $\frac{6.23 \text{ mol}}{6.23 \text{ mol}}$ = 1.00 x 4 = 4 |

For C$_5$H$_4$, EFM = 64.0 g/mol.

To calculate x, the number of formula units per molecule:
$$x = \left(\frac{MM}{EFM}\right) = \left(\frac{128}{64.0}\right) = 2.00 \text{ or } 2$$

The molecular formula is (C$_5$H$_4$)$_2$ or C$_{10}$H$_8$.

7.65 (a) 55.9 g Fe (b) 63.6 g Cu (c) 108 g Ag (d) 24.3 g Mg (e) 52.0 g Cr
 (f) 24.3 g Mg

(a) Find: mass of metal (g) = ?
 Given: 1.00 mol Fe(NO$_3$)$_3$
 Known: MM-Fe = 55.9 g/mol
  1 mol Fe/mol Fe(NO$_3$)$_3$
 Solution: mass (g) = $\left(\frac{55.9 \text{ g Fe}}{1 \text{ mol Fe}}\right)\left(\frac{1 \text{ mol Fe}}{1 \text{ mol Fe(NO}_3)_3}\right)$(1.00 mol Fe(NO$_3$)$_3$) = 55.9 g Fe

(b) Find: mass of metal (g) = ?
 Given: 1.00 mol CuCl$_2$
 Known: MM-Cu = 63.6 g/mol
  1 mol Cu/mol CuCl$_2$
 Solution: mass (g) = $\left(\frac{63.6 \text{ g Cu}}{1 \text{ mol Cu}}\right)\left(\frac{1 \text{ mol Cu}}{1 \text{ mol CuCl}_2}\right)$(1.00 mol CuCl$_2$) = 63.6 g Cu

(c) Find: mass of metal (g) = ?
 Given: 1.00 mol AgCl
 Known: MM-Ag = 108 g/mol
  1 mol Ag/mol AgCl
 Solution: mass (g) = $\left(\frac{108 \text{ g Ag}}{1 \text{ mol Ag}}\right)\left(\frac{1 \text{ mol Ag}}{1 \text{ mol AgCl}}\right)$(1.00 mol AgCl) = 108 g Ag

(d) Find: mass of metal (g) = ?

Given: 1.00 mol Mg(OH)$_2$

Known: MM-Mg = 24.3 g/mol

1 mol Mg/mol Mg(OH)$_2$

Solution: mass (g) = ($\frac{24.3 \text{ g Mg}}{1 \text{ mol Mg}}$)($\frac{1 \text{ mol Mg}}{1 \text{ mol Mg(OH)}_2}$)(1.00 mol Mg(OH)$_2$) = 24.3 g Mg

(e) Find: mass of metal (g) = ?

Given: 1.00 mol CrO$_3$

Known: MM-Cr = 52.0 g/mol

1 mol Cr/mol CrO$_3$

Solution: mass (g) = ($\frac{52.0 \text{ g Cr}}{1 \text{ mol Cr}}$)($\frac{1 \text{ mol Cr}}{1 \text{ mol CrO}_3}$)(1.00 mol CrO$_3$) = 52.0 g Cr

(f) Find: mass of metal (g) = ?

Given: 1.00 mol Mg(ClO$_4$)$_2$

Known: MM-Mg = 24.3 g/mol

1 mol Mg/mol Mg(ClO$_4$)$_2$

Solution: mass (g) = ($\frac{24.3 \text{ g Mg}}{1 \text{ mol Mg}}$)($\frac{1 \text{ mol Mg}}{1 \text{ mol Mg(ClO}_4)_2}$)(1.00 mol Mg(ClO$_4$)$_2$) = 24.3 g Mg

7.67 (a) 0.816 g (b) 0.835 g (c) 0.500 g (d) 0.441 g

(a) Find: mass of chlorine (g) = ?

Given: 1.00 g Cl$_2$O

Known: 71.0 g Cl/87.0 g Cl$_2$O (from molar masses of the elements);

or 2 mol Cl/mol Cl$_2$O, MM-Cl$_2$O = 87.0 g/mol, MM-Cl = 35.5 g/mol;

or 81.6 g Cl/100 g Cl$_2$O from percentage composition calculated in 7.33a

Solution: mass (g) = ($\frac{71.0 \text{ g Cl}}{87.0 \text{ g Cl}_2\text{O}}$)(1.00 g Cl$_2$O) = 0.816 g Cl

or

$$\text{mass (g)} = \left(\frac{35.5 \text{ g Cl}}{1 \text{ mol Cl}}\right)\left(\frac{2 \text{ mol Cl}}{1 \text{ mol Cl}_2\text{O}}\right)\left(\frac{1 \text{ mol Cl}_2\text{O}}{87.0 \text{ g Cl}_2\text{O}}\right)(1.00 \text{ g Cl}_2\text{O}) = 0.816 \text{ g Cl}$$

or

$$\text{mass (g)} = \left(\frac{81.6 \text{ g Cl}}{100 \text{ g Cl}_2\text{O}}\right)(1.00 \text{ g Cl}_2\text{O}) = 0.816 \text{ g Cl}$$

(b) Find: mass of chlorine (g) = ?

Given: 1.00 g CH_2Cl_2

Known: 83.5 g Cl/100 g CH_2Cl_2 from percentage composition calculated in 7.33c

Solution: $\text{mass (g)} = \left(\frac{83.5 \text{ g Cl}}{100 \text{ g CH}_2\text{Cl}_2}\right)(1.00 \text{ g CH}_2\text{Cl}_2) = 0.835 \text{ g Cl}$

(c) Find: mass of chlorine (g) = ?

Given: 1.00 g $C_{14}H_9Cl_5$

Known: 50.0 g Cl/100 g $C_{14}H_9Cl_5$ from percentage composition calculated in 7.33d.

Solution: $\text{mass (g)} = \left(\frac{50.0 \text{ g Cl}}{100 \text{ g C}_{14}\text{H}_9\text{Cl}_5}\right)(1.00 \text{ g C}_{14}\text{H}_9\text{Cl}_5) = 0.500 \text{ g Cl}$

(d) Find: mass of chlorine (g) = ?

Given: 1.00 g $C_{12}H_4Cl_4O_2$

Known: 44.1 g Cl/100 g $C_{12}H_4Cl_4O_2$ from percentage composition calculated in 7.33e.

Solution: $\text{mass (g)} = \left(\frac{44.1 \text{ g Cl}}{100 \text{ g C}_{12}\text{H}_4\text{Cl}_4\text{O}_2}\right)(1.00 \text{ g C}_{12}\text{H}_4\text{Cl}_4\text{O}_2) = 0.441 \text{ g Cl}$

7.69 $Na_2Al_2Si_3O_{10} \cdot 2H_2O$

Element	Step 1: mass, g	Step 2: number of moles	Step 3: simplest whole-number mole ratio	Step 4: formula
sodium	12.1 g	$\frac{12.1 \text{ g}}{23.0 \text{ g/mol}} = 0.526$ mol	$\frac{0.526 \text{ mol}}{0.526 \text{ mol}} = 1.00 \times 2$	$Na_2Al_2Si_3O_{10} \cdot 2H_2O$
aluminum	14.2 g	$\frac{14.2 \text{ g}}{27.0 \text{ g/mol}} = 0.526$ mol	$\frac{0.526 \text{ mol}}{0.526 \text{ mol}} = 1.00 \times 2$	
silicon	22.1	$\frac{22.1 \text{ g}}{28.1 \text{ g/mol}} = 0.786$ mol	$\frac{0.786 \text{ mol}}{0.526 \text{ mol}} = 1.49 \times 2$	

66   Student's Solutions Manual

oxygen	42.1 g	$\frac{42.1 \text{ g}}{16.0 \text{ g/mol}} = 2.63$ mol	$\frac{2.63 \text{ mol}}{0.526 \text{ mol}} = 5.00 \times 2$	
water	9.48 g	$\frac{9.48 \text{ g}}{18.0 \text{ g/mol}} = 0.527$ mol	$\frac{0.527 \text{ mol}}{0.526 \text{ mol}} = 1.00 \times 2$	

7.71 142 g Cl

Find: mass of chlorine (g) = ?
Given: 1.00 mol CCl$_4$
Known: AM-Cl = 35.5 g/mol and $\frac{4 \text{ mol Cl}}{1 \text{ mol CCl}_4}$
Solution: mass (g) = $(\frac{35.5 \text{ g Cl}}{1 \text{ mol Cl}})(\frac{4 \text{ mol Cl}}{1 \text{ mol CCl}_4})(1.00 \text{ mol CCl}_4)$ = 142 g Cl

7.73 1.64×10^7 g

Find: mass (g) = ?
Given: 5.00×10^{28} atoms Au
Known: 6.02×10^{23} atoms/mol
             AM-Au = 197 g/mol
Solution: mass (g) = $(\frac{197 \text{ g}}{1 \text{ mol}})(\frac{1 \text{ mol}}{6.02 \times 10^{23} \text{ atoms}})(5.00 \times 10^{28} \text{ atoms})$ = 1.64×10^7 g

7.75

compound	mass, g	number of moles	number of molecules
C$_{14}$H$_9$Cl$_5$	88.8 (a)	0.250	1.51×10^{23} (b)
C$_{10}$H$_8$	10.0	0.0781 (c)	4.70×10^{22} (d)
NO$_2$	95.7 (f)	2.08 (e)	1.25×10^{24}
C$_2$H$_6$O	2.68×10^{-21} (h)	5.81×10^{-23} (g)	35

Chapter 7: Composition and Formulas of Compounds 67

(a) Find: mass of $C_{14}H_9Cl_5$ (g) = ?

Given: 0.250 mol $C_{14}H_9Cl_5$

Known: MM-$C_{14}H_9Cl_5$ = 355 g/mol

Solution: mass (g) = ($\frac{355 \text{ g}}{1 \text{ mol}}$)(0.250 mol) = 88.8 g

(b) Find: number of molecules = ?

Given: 0.250 mol $C_{14}H_9Cl_5$

Known: 6.02 x 10^{23} molecules/mol

Solution: molecules = ($\frac{6.02 \times 10^{23} \text{ molecules}}{1 \text{ mol}}$)(0.250 mol) = 1.51 x 10^{23} molecules

(c) Find: number of moles (mol) = ?

Given: 10.0 g

Known: MM-$C_{10}H_8$ = 128 g/mol

Solution: mol = ($\frac{1 \text{ mol}}{128 \text{ g}}$)(10.0 g) = 0.0781 mol

(d) Find: number of molecules = ?

Given: 0.0781 mol

Known: 6.02 x 10^{23} molecules/mol

Solution: molecules = ($\frac{6.02 \times 10^{23} \text{ molecules}}{1 \text{ mol}}$)(0.0781 mol) = 4.70 x 10^{22} molecules

(e) Find: number of moles (mol) = ?

Given: 1.25 x 10^{24} molecules

Known: 6.02 x 10^{23} molecules/mol

Solution: mol = $\frac{1 \text{ mol}}{6.02 \times 10^{23} \text{ molecules}}$)(1.25 x 10^{24} molecules) = 2.08 mol

(f) Find: mass (g) = ?

Given: 2.08 mol

Known: MM-NO_2 = 46.0 g/mol

Solution: mass (g) = ($\frac{46.0 \text{ g}}{1 \text{ mol}}$)(2.08 mol) = 95.7 g

(g) Find: number of moles (mol) = ?

Given: 35 molecules

Known: 6.02 x 10²³ molecules/mol

Solution: mol = $(\frac{1 \text{ mol}}{6.02 \times 10^{23} \text{ molecules}})(35 \text{ molecules}) = 5.81 \times 10^{-23}$ mol

(h) Find: mass (g) = ?

Given: 5.81 x 10⁻²³ mol

Known: MM-C₂H₆O = 46.1 g/mol

Solution: mass (g) = $(\frac{46.1 \text{ g}}{1 \text{ mol}})(5.81 \times 10^{-23} \text{ mol}) = 2.68 \times 10^{-21}$ g

7.77 1.50 g C

Find: mass C (g) = ?

Given: 4.00 g S

Known: Equal moles contain equal numbers of atoms, or 1 mol S ≅ 1 mol C, or 1 mol S/1 mol C

Solution: Calculate moles of S in 4.00 g of S. This equals moles of C that contain the same number of atoms as there are in 4.00 g of S.

mass C (g) = $(\frac{12.0 \text{ g C}}{1 \text{ mol C}})(\frac{1 \text{ mol C}}{1 \text{ mol S}})(\frac{1 \text{ mol S}}{32.1 \text{ g S}})(4.00 \text{ g S}) = 1.50$ g C

7.79 (a) 40.1% Ca, 12.0% C, 48.0% O (b) 41.7% Mg, 54.9% O, 3.46% H

(a) % Ca = $(\frac{40.1 \text{ g}}{100. \text{ g}})(100) = 40.1\%$, % C = $(\frac{12.0 \text{ g}}{100. \text{ g}})(100) = 12.0\%$,

% O = $(\frac{(3)(16.0 \text{ g})}{100. \text{ g}})(100) = 48.0\%$

(b) %Mg = $(\frac{24.3 \text{ g}}{58.3 \text{ g}})(100) = 41.7\%$, % O = $(\frac{(2)(16.0 \text{ g})}{58.3 \text{ g}})(100) = 54.9\%$,

%H g= $(\frac{(2)(1.01 \text{ g})}{58.3 \text{ g}})(100) = 3.46\%$

Chapter 8:
The Structure of Compounds. Chemical Bonds

8.1 ionic and covalent

8.3 An ionic bond is a bond (attraction) between ions of opposite charge in a compound. A covalent bond is a bond (attraction) between atoms in a compound that results from sharing a pair of electrons. The feature of ionic bonding that helps to distinguish it from a covalent bond is that it is multidirectional.

8.5 Size and relative charge of the ions affect the packing arrangement in an ionic crystal.

8.7 high melting points, solids at room temperature, hard, rigid, shatter easily upon impact, nonconductors of electricity in the solid state, conductors of electricity when dissolved in water and when fused (molten)

Physical properties can be observed without changing the chemical identity of a substance.

8.9 covalent bonding

8.11 ionic: (a), (c), and (f); covalent: (b), (d), and (e)

Typically, ionic bonding is observed between metals and nonmetals. Covalent bonding is observed between two or more nonmetals.

8.13 endothermic

The fact that light is required indicates that this reaction is endothermic, as light is a form of energy. Furthermore, breaking bonds always requires energy and therefore is always endothermic.

8.15 (a) CCl_4 (b) CCl_4

The species having the bonds is lower in energy. Remember that energy must be supplied to break a bond. Therefore C and 4 Cl (4 broken bonds) or Cl and CCl_3 (1 broken bond) are both higher energy states than CCl_4.

8.17 Lewis formula

8.19 (a) $\left[\begin{array}{c} H \\ | \\ H-N-H \\ | \\ H \end{array} \right]^+$ $\left[:\ddot{\underset{..}{Cl}}: \right]^-$ (b) $\begin{array}{c} H \\ | \\ H-As: \\ | \\ H \end{array}$ (c) $\begin{array}{c} :\ddot{Cl}: \\ | \\ :\ddot{Cl}-N: \\ | \\ :\ddot{Cl}: \end{array}$ (d) $K^+ \left[:\ddot{\underset{..}{O}}-H \right]^-$

(e) $:\ddot{\underset{..}{O}}-\ddot{\underset{..}{Se}}=\ddot{\underset{..}{O}}:$ (f) $Mg^{2+}\ 2\left[:\ddot{\underset{..}{Br}}: \right]^-$

(a) NH$_4$Cl is an ionic compound. Each ion must be represented separately. Both ions have Lewis formulas. However, the Lewis formula for monatomic anions like Cl$^-$ is simply the Lewis symbol that you learned in Chapter 6. Follow the steps to write the Lewis formula for the cation. Because the chloride anion has a -1 charge, the cation is +1.

Step 1: Calculate the total number of valence electrons for the atoms.

nitrogen (Group 5A):	5 e$^-$ x 1 atom	= 5 e$^-$
hydrogen (atomic number = 1)	1 e$^-$ x 4 atoms	= 4 e$^-$
+1 charge		= -1 e$^-$
	total	= 8 e$^-$

Step 2: Predict the number of bonds for each atom.

3 bonds for nitrogen (8 - 5 = 3)

1 bond for hydrogen

Step 3: Arrange the atoms for symmetry and connect with single bonds:

```
      H
      |
  H—N—H
      |
      H
```

Step 4: Add bonds to agree with Step 2.

Since the number of bonds for nitrogen exceeds the number predicted, go on to Step 5.

Step 5: Add pairs of dots for nonbonding electrons.

Nitrogen already has the use of an octet of electrons, so there are no nonbonding pairs of electrons.

Step 6: Check the number of valence electrons.

There are eight valence electrons, so it checks. Remember the charge of the ion.

$$\left[\begin{array}{c} \text{H} \\ | \\ \text{H}-\text{N}-\text{H} \\ | \\ \text{H} \end{array}\right]^{+} \left[:\ddot{\underset{..}{\text{Cl}}}:\right]^{-}$$

(b) Step 1: arsenic 5 e⁻ x 1 atom = 5 e⁻

     hydrogen 1 e⁻ x 3 atoms = 3 e⁻

     total   8 e⁻

  Step 2: Predict 3 bonds for arsenic and 1 bond each for hydrogen.

  Step 3: Arrange the atoms for symmetry and connect the atoms with single bonds:

$$\begin{array}{c} \text{H} \\ | \\ \text{H}-\text{As} \\ | \\ \text{H} \end{array}$$

  Step 4: The number of bonds is correct for all the atoms.

  Step 5: Add 1 pair of electrons to arsenic:

$$\begin{array}{c} \text{H} \\ | \\ \text{H}-\text{As}: \\ | \\ \text{H} \end{array}$$

  Step 6: Check valence electrons. Because the formula has 8 e⁻, it checks.

(c) Step 1: nitrogen 5 e⁻ x 1 atom = 5 e⁻

     chlorine   7 e⁻ x 3 atoms = 21 e⁻

     total   26 e⁻

  Step 2: Predict 3 bonds for nitrogen and 1 bond each for chlorine.

  Step 3: Arrange the atoms for symmetry and connect the atoms with single bonds:

$$\begin{array}{c} \text{Cl} \\ | \\ \text{Cl}-\text{N} \\ | \\ \text{Cl} \end{array}$$

  Step 4: The number of bonds is correct for all the atoms.

  Step 5: Add 1 pair of electrons to nitrogen and 3 pairs to each chlorine:

$$\begin{array}{c} \overset{..}{:}\overset{..}{\text{Cl}}: \\ | \\ :\overset{..}{\underset{..}{\text{Cl}}}-\text{N}: \\ | \\ :\overset{..}{\underset{..}{\text{Cl}}}: \end{array}$$

Step 6:   Check valence electrons. Because the formula has 26 e⁻, it checks.

(d) Step 1:   potassium ion: K⁺

hydroxide ion: 6 e⁻ (O) + 1 e⁻ (H) + 1 e⁻ (charge) = 8 e⁻

Step 2:   Predict 2 bonds for oxygen and 1 bond for hydrogen in the hydroxide ion.

Step 3:   Arrange the atoms for symmetry and connect the atoms with single bonds:

$$\left[\; \text{O}-\text{H} \;\right]$$

Step 4:   Hydrogen can only form one bond, so oxygen cannot have two bonds in this ion.

Step 5:   Add 3 pairs of electrons to oxygen:

$$\left[\; :\overset{..}{\underset{..}{\text{O}}}-\text{H} \;\right]^{-}$$

Step 6:   The number of valence electrons checks.

$$\text{K}^{+}\left[\; :\overset{..}{\underset{..}{\text{O}}}-\text{H} \;\right]^{-}$$

(e) Step 1:   6 e⁻ (Se) + 12 e⁻ (2 O) = 18 e⁻

Step 2:   Predict 2 bonds for selenium and 2 bonds each for oxygen.

Step 3:   Arrange the atoms for symmetry and connect the atoms with single bonds:

O—Se—O

Step 4:   Selenium has the correct number of bonds, but neither of the oxygens has the correct number of bonds. Go to Step 5.

Step 5:   Since 4 electrons were used in forming two covalent bonds, there are 18 - 4 or 14 electrons available to distribute in seven pairs of electrons around the atoms. Give three pairs of electrons to each oxygen atom and one pair of electrons to the selenium atom:

$$:\ddot{\underset{..}{O}}-\ddot{Se}-\ddot{\underset{..}{O}}:$$

Step 6: Check valence electrons. All 18 e⁻ are accounted for, but the selenium does *not* have an octet of electrons. It will be necessary to form a double bond by moving a nonbonded pair of electrons from one of the oxygens to form a covalent bond with selenium. Since you could take the pair of electrons from either oxygen atom, resonance is possible.

$$:\ddot{\underset{..}{O}}-\ddot{Se}=\ddot{O}: \longleftrightarrow :\ddot{O}=\ddot{Se}-\ddot{\underset{..}{O}}:$$

(f) This is an ionic compound in which each of the ions is monatomic. Use the simple dot symbols learned in Chapter 6:

$$Mg^{2+} \quad 2\left[:\ddot{\underset{..}{Br}}:\right]^{-}$$

8.21

$$Na^{+} \left[:\overset{..}{\underset{..}{O}}=\overset{\overset{:\ddot{O}-H}{|}}{C}-\ddot{\underset{..}{O}}: \longleftrightarrow :\ddot{\underset{..}{O}}-\overset{\overset{:\ddot{O}-H}{|}}{C}=\ddot{O}: \right]^{-}$$

Sodium is a simple ion. There are 24 electrons [4 e⁻ (C) + 18 e⁻ (3 O) + 1 e⁻ (H) + 1 e⁻ (charge must be 1- because the charge on the sodium ion is 1+)]. Since carbon tends to form the most bonds (4), it is used as the central atom. Hydrogen is bonded to one of the oxygen atoms (any one of them!) before distributing the remaining electrons. The only way to satisfy the octet rule for all the atoms is to form one double bond between carbon and one of the two oxygen atoms that is not bonded to hydrogen. The source of resonance lies in the fact that you can form the double bond in either of two places.

8.23 $:\ddot{\text{O}}-\ddot{\text{S}}=\ddot{\text{O}}: \longleftrightarrow :\ddot{\text{O}}=\ddot{\text{S}}-\ddot{\text{O}}:$

Since sulfur and oxygen are both in Group 6A, there are 3 x 6 e⁻ or 18 e⁻. Sulfur is placed in the middle. In the absence of directions otherwise, assume that a symmetrical structure will be formed. All three atoms want to form two bonds each, but this is not possible. If 14 nonbonded electrons are distributed around the atoms, one of the three atoms only gets the use of 6 electrons. In order for each atom to satisfy the octet rule, there must be one double bond. However, this double bond can be drawn to either oxygen atom, so resonance is possible.

8.25 O_2 has unpaired electrons, which make it paramagnetic. N_2 does not have unpaired electrons.

Oxygen in known to be paramagnetic by experimental evidence. Therefore, it must contain unpaired electrons.

8.27 Fluorine is the most electronegative element.

8.29 polar: **(a)** and **(c)**   nonpolar: **(b)** and **(d)**

Polar bonds form between atoms of different electronegativity. The difference in electronegativity for the four bonds in this problem is (a) 4.0 (F) - 2.2 (H) = 1.8, polar; (b) 2.6 (C) - 2.2 (H) = 0.4, essentially nonpolar; (c) 3.4 (O) - 2.2 (H) = 1.2, polar; and (d) 3.2 (Cl) - 3.0 (N) = 0.2, essentially nonpolar.

8.31 Because the two different atoms joined together have almost the same electronegativity (difference in electronegativity = 0.2), and therefore essentially the same attraction for shared electrons, the bond is nonpolar.

8.33 (a) H-F (b) equally polar (c) O-H (d) C-O (e) P-Cl

To determine which bond is more polar in each pair, you must determine the electronegativity difference for each bond in the pair. The bond having the larger electronegativity difference is more polar. (a) H-F = 4.0 - 2.2 = 1.8, H-Cl = 3.2 - 2.2 = 1.0; (b) S-O = 3.4 - 2.6 = 0.8, C-O = 3.4 - 2.6 = 0.8; (c) N-H = 3.0 - 2.2 = 0.8, O-H = 3.4 - 2.2 = 1.2; (d) N-O = 3.4 - 3.0 = 0.4, C-O = 3.4 - 2.6 = 0.8; (e) P-Cl = 3.2 - 2.2 = 1.0, C-Cl = 3.2 - 2.6 = 0.6.

8.35 (a) tetrahedral (b) linear (c) linear (d) trigonal

To identify the shape of these molecules, first draw the Lewis formulas for them and then count the number of pairs of electrons to the central atom.

(a) There are four sets around C, so the molecule is tetrahedral. (b) There are two sets of electrons around each carbon atom (multiple bonds count as one set of electrons for purposes of molecular shape), so the molecule is linear. (c) There are two sets of electrons around C, so the molecule is linear. (d) There are three sets of electrons around C, so the molecule is trigonal.

8.37 H—C≡N:; bond angles are 180°; shape is linear

There are two sets of electrons around carbon, because triple bonds are counted as one set. Therefore the molecule has 180° angles and is linear.

8.39 not polar; carbon and sulfur have the same electronegativity, so there are no polar bonds; a molecule cannot be polar if there are no polar bonds, regardless of its shape.

8.41 The Lewis formulas, below, show that the S in sulfur dioxide has three sets of electrons, whereas the C in carbon dioxide has two sets of electrons. In carbon dioxide, there are two equal dipoles which exactly cancel because of the linear shape of the molecule. Therefore CO_2 is nonpolar. The extra set of electrons on the S in sulfur dioxide yields a shape with bond angles of 120°. There are two dipoles, but they do not cancel one another because of the bond angles. Therefore, SO_2 is polar.

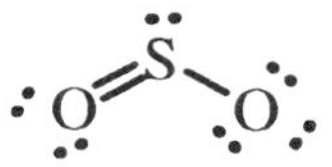

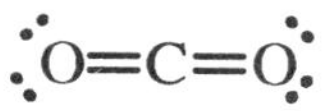

8.43 Propane is not polar because it does not contain polar bonds.

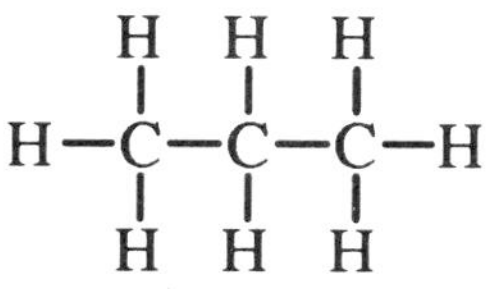

```
    H   H   H
    |   |   |
H – C – C – C – H
    |   |   |
    H   H   H
```

8.45 The Lewis formulas drawn below are not resonance structures because they do differ in the placement of electrons as well as the placement of atoms.

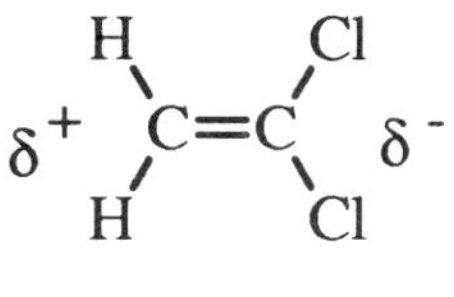

δ^+ C=C δ^- (H, H / Cl, Cl) — polar

C=C (δ^+ H, H / Cl, Cl δ^-) — polar

C=C (H, Cl / Cl, H) — nonpolar

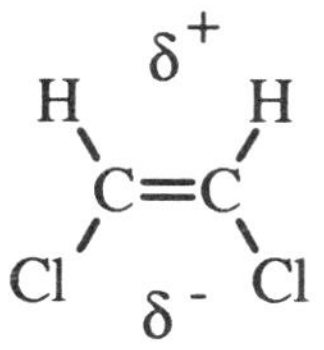

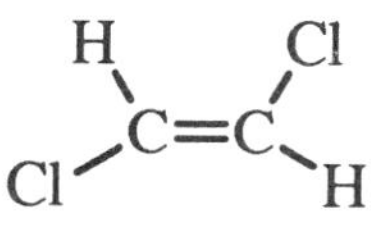

8.47 (a)

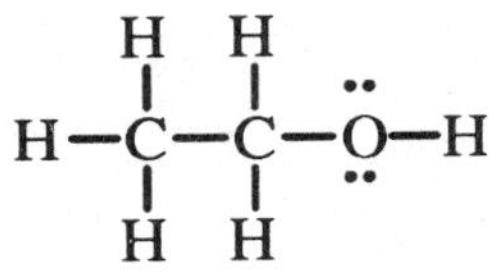

H–C(H)(H)–C(H)(H)–Ö–H and H–C(H)(H)–Ö–C(H)(H)–H

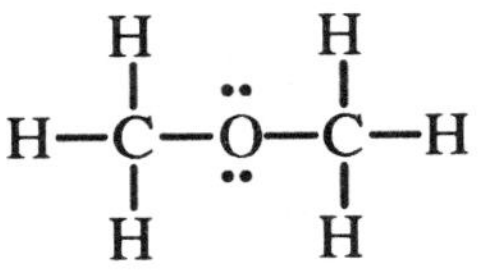

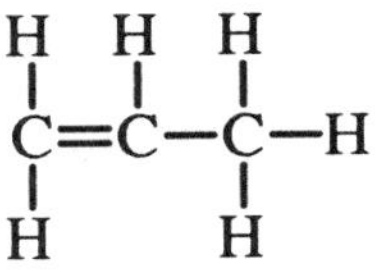

(b) H(H)C=C(H)–C(H)(H)–H and cyclopropane: H–C(H)–C(H)–H with CH₂ bridge

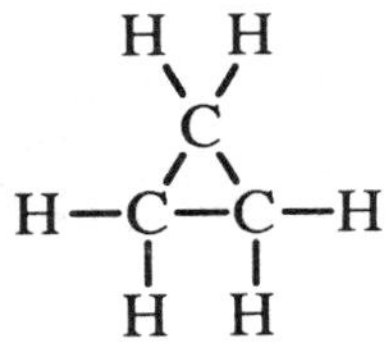

8.49

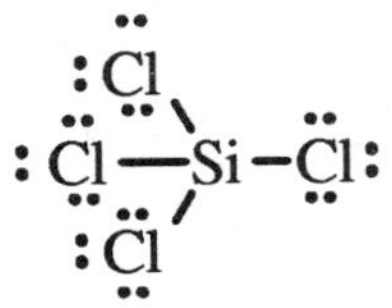

:Cl–Si(–Cl:)(–Cl:)–Cl: bond angles = 109°; molecule is tetrahedral; nonpolar because the bond dipoles cancel one another.

8.51

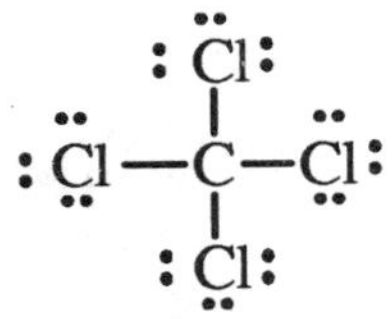

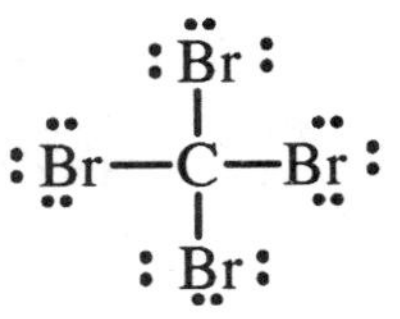

:Cl–C(:Cl:)(:Cl:)–Cl: :Br–C(:Br:)(:Br:)–Br: Both compounds are nonpolar because the tetrahedral shape results in the cancellation of the bond dipoles.

Chapter 9:
Names and Formulas of Inorganic Compounds

9.1 (a) ammonia (b) baking soda (c) water (d) lye

These common names are listed in Table 9.1.

9.3 marble

9.5 (a) -3 (b) +4 (c) +3 (d) +3

	(a) NH_3	(b) NO_2	(c) NCl_3	(d) HNO_2
formula:				
Step 1:	+1	-2	-1	+1 -2
Step 2:	3(+1)	2(-2)	3(-1)	(+1) 2(-2)
Step 3:	N + 3 = 0	N + (-4) = 0	N + (-3) = 0	(+1) + N + (-4) = 0
	N = -3	N = +4	N = +3	N = +3

9.7 (a) +7 (b) +5 (c) +3 (d) +1

(a) $HClO_4$: (+1) + Cl + 4(-2) = 0, Cl = +7   (b) $HClO_3$: (+1) + Cl + 3(-2) = 0, Cl = +5

(c) HClO$_2$: (+1) + Cl + 2(-2) = 0, Cl = +3   (d) HClO: (+1) + Cl + (-2) = 0, Cl = +1

9.9 (a) -2   (b) -1   (c) -2   (d) -2

(a), (c), and (d) The oxidation number of oxygen is usually -2 in compounds (Rule 3). (b) This compound is hydrogen peroxide. If the oxidation number of oxygen were -2, then the oxidation number of hydrogen would have to be +2 each. This is never observed. If we assume that the oxidation number of hydrogen is +1, then the oxidation number of oxygen must be -1, a condition observed in peroxides.

9.11 (a) S = +4   (b) P = -3   (c) Cl = +7   (d) Se = -2

(a) H$_2$SO$_3$: 2(+1) + S + 3(-2) = 0, S = +4   (c) Cl$_2$O$_7$: 2 Cl + 7(-2) = 0, Cl = +7
(b) PH$_3$: P + 3(+1) = 0, P = -3   (d) H$_2$Se: 2(+1) + Se = 0, Se = -2

9.13 (a) I = +7, O = -2   (b) N = -3, H = +1   (c) As = +5, O = -2   (d) Mn = +7, O = -2

(a) IO$_4^-$: I + 4(-2) = -1, I = +7   (c) AsO$_4^{3-}$: As + 4(-2) = -3, As = +5
(b) NH$_4^+$: N + 4(+1) = +1, N = -3   (d) MnO$_4^-$: Mn + 4(-2) = -1, Mn = +7

9.15 (a) potassium iodide   (b) tetraphosphorus decoxide   (c) iron(II) oxide   (d) tin(IV) sulfide
(e) magnesium oxide

(a) ionic: Name the cation first followed by the anion ending in -ide. (b) covalent: Name the elements using prefixes. (c) ionic: Name the cation first followed by the anion ending in -ide. Since oxygen is -2, iron must be +2. Therefore, the cation is Fe^{2+} or iron(II) ion. (d) probably covalent, but named as if ionic: Name the cation first followed by the anion ending in -ide. Since there are two sulfide ions at -2 each, the apparent charge on tin is +4. Therefore, the cation is named tin(IV). Prefixes are not used. (e) ionic: Name the cation first followed by the anion ending in -ide.

9.17 (a) iron(III) ion   (b) mercury(I) ion   (c) tin(IV) ion   (d) copper(I) ion

Systematic names use Roman numerals to identify the oxidation numbers of metals that can have more than one oxidation state in compounds.

9.19 (a) copper(II) chloride   (b) iron(II) nitride   (c) tin(IV) iodide   (d) copper(I) sulfide

(a) Cu + 2(-1) = 0, Cu = +2, copper(II) chloride   (b) 3 Fe + 2(-3) = 0, Fe = +2, iron(II) nitride
(c) Sn + 4(-1) = 0, Sn = +4, tin(IV) iodide   (d) 2 Cu + (-2) = 0, Cu = +1, copper(I) sulfide

9.21 (a) Al_2S_3 (b) $PbBr_4$ (c) $Cu(CN)_2$ (d) $SnCl_2$ (e) Mg_3N_2

(a) $Al^{3+} + Al^{3+} + S^{2-} + S^{2-} + S^{2-} = 0$ (b) $Pb^{4+} + 4\ Br^- = 0$ (c) $Cu^{2+} + 2\ CN^- = 0$
(d) $Sn^{2+} + 2\ Cl^- = 0$ (e) $3\ Mg^{2+} + 2\ N^{3-} = 0$

9.23 (a) potassium cyanide   (b) magnesium hydroxide   (c) copper(II) hydroxide   (d) iron(III) hydroxide

All of these compounds are ionic, so the name consists of the name of the cation followed by the name of the anion. For $Cu(OH)_2$, the charge or oxidation number of copper can be figured out by working backwards from 2 hydroxide ions at -1 each. The copper must be +2 and the compound is named copper(II) hydroxide. For Fe(OH)$_3$, similar reasoning indicates Fe^{3+}, and the compound is named iron(III) hydroxide.

9.25 (a) Sn(OH)$_4$ (b) NH$_4$Br   (c) Fe(OH)$_2$

(a) $+4 + x(-1) = 0$, $x = 4$, Sn(OH)$_4$ (b) $+1 + (-1) = 0$, NH$_4$Br   (c) $+2 + x(-1)$, $x = 2$, Fe(OH)$_2$

Remember that when more than one polyatomic ion must be indicated in a formula, the polyatomic ion is enclosed in parentheses and the subscript is placed after the parentheses. SnOH$_4$ describes a compound containing one atom of tin, one atom of oxygen, and four atoms of hydrogen. It is not the same as Sn(OH)$_4$, which has one atom of tin, four atoms of oxygen, and four atoms of hydrogen.

9.27 binary acids: (b) HBr(aq), (d) H$_2$S(aq)

Binary acids can be recognized by three characteristics: two elements only, the first element listed is hydrogen, and the material is dissolved in water (aq).

9.29 (a) HI(aq)   (b) HBr(aq)   (c) H$_2$Se(aq)

The number of hydrogen atoms must balance the oxidation number of the second element. Iodine and bromine have a -1 oxidation number in binary compounds with elements other than oxygen. Therefore, they combine with one atom of hydrogen to form binary acids. Selenium is in Group 6A, so it has an oxidation number of -2 and combines with two hydrogen atoms to form a binary acid.

Chapter 9: Names and Formulas of Inorganic Compounds   83

9.31 (a) arsenic acid (b) hypobromous acid (c) selenic acid (d) selenous acid (e) iodic acid (f) arsenous acid

(a) arsenic acid (related to H_3PO_4, phosphoric acid) (b) hypobromous acid (related to HClO, hypochlorous acid) (c) selenic acid (related to H_2SO_4, sulfuric acid) (d) selenous acid (related to H_2SO_3, sulfurous acid) (e) iodic acid (related to $HClO_3$, chloric acid) (f) arsenous acid (related to H_3PO_3, phosphorous acid)

9.33 (a) HNO_3 (b) $HClO_2$ (c) HIO_4 (d) H_2SO_3

(a) HNO_3 (memorized) (b) $HClO_2$ (one less oxygen than chloric acid) (c) HIO_4 (one more oxygen than iodic acid, which is related to chloric acid) (d) H_2SO_3 (one less oxygen atom than sulfuric acid)

9.35 (a) sodium hypochlorite (b) calcium sulfite (c) potassium nitrate (d) sodium nitrite

(a) cation: Na^+ is sodium ion; anion: OCl^- is related to HOCl, hypochlorous acid, hypo-ous/hypo-ite, therefore hypochlorite ion; name: sodium hypochlorite.

(b) cation: calcium ion; anion: H_2SO_3/sulfurous acid, ous/ite, sulfite ion; name: calcium sulfite

(c) cation: potassium ion; anion: HNO_3/nitric acid, ic/ate, nitrate ion; name: potassium nitrate

(d) cation: sodium ion; anion: HNO_2/nitrous acid, ous/ite, nitrite ion; name: sodium nitrite.

9.37 (a) copper(II) nitrate (b) tin(II) sulfate (c) copper(II) phosphate

(a) cation: Cu^{2+} from two nitrates at -1 each; anion: HNO_3/nitric acid, ic/ate, nitrate ion; name: copper(II) nitrate (b) cation: Sn^{2+} from one sulfate at -2; anion: H_2SO_4/sulfuric acid, ic/ate, sulfate ion; name: tin(II) sulfate (c) cation: Cu^{2+} from balance of charge against two phosphates at -3 each; anion: H_3PO_4/phosphoric acid, ic/ate, phosphate ion; name: copper(II) phosphate.

9.39 nitrous acid, HNO_2

The -ous acid goes with the -ite anion.

9.41 (a) potassium bisulfite (b) copper(II) bisulfate (c) sodium biselenate

(a) cation: potassium ion; anion: one less oxygen than bisulfate, bisulfite ion; name: potassium bisulfite (b) cation: copper(II) ion from two bisulfates at -1 each; anion: related to sulfate ion, bisulfate ion; name: copper(II) bisulfate (c) cation: sodium ion; anion: related to HSO_4^- (bisulfate ion), biselenate ion; name: sodium biselenate.

9.43 (a) NO_3^- (b) ClO_4^- (c) NO_2^- (d) IO_3^- (e) SO_4^{2-} (f) ClO_2^- (g) ClO^- (h) BrO_3^-

(a) related to nitric acid, HNO_3 (b) related to perchloric acid, $HClO_4$ (c) related to nitrous acid, HNO_2 (d) related to iodic acid, HIO_3 (e) related to sulfuric acid, H_2SO_4 (f) related to chlorous acid, $HClO_2$ (g) related to hypochlorous acid, $HClO$ (h) related to bromic acid, $HBrO$

9.45 (a) MnO_4^- (b) ClO_3^- (c) $Cr_2O_7^{2-}$ (d) CrO_4^{2-} (e) PO_4^{3-}

(a) memorized (b) related to chloric acid, $HClO_3$ (c) memorized (d) memorized

(e) related to phosphoric acid, H3PO4

9.47 (a) Fe$_2$(SO$_4$)$_3$ (b) Cu$_3$PO$_4$ (c) Cr$_2$(SO$_4$)$_3$ (d) Sn(ClO$_2$)$_2$ (e) Na$_2$CO$_3$

(a) cation: Fe^{3+}; anion: ate/ic, sulfuric acid/H$_2$SO$_4$, SO$_4^{2-}$; 2(+3) + 3(-2) = 0

(b) cation: Cu$^+$; anion: ate/ic, phosphoric acid/H$_3$PO$_4$, PO$_4^{3-}$, 3(+1) + (-3) = 0

(c) cation: Cr^{3+}, anion: ate/ic, sulfuric acid/H$_2$SO$_4$, SO$_4^{2-}$; 2(+3) + 3(-2) = 0

(d) cation: Sn^{2+}, anion: ite/ous, chlorous acid/HClO$_2$, ClO$_2^-$, (+2) + 2(-1) = 0

(e) cation: Na$^+$, anion: ate/ic, carbonic acid/H$_2$CO$_3$, CO$_3^{2-}$, 2(+1) + (-2) = 0

9.49 (a) KMnO$_4$ (b) K$_2$Cr$_2$O$_7$ (c) Na$_2$CrO$_4$ (d) NaClO$_3$

(a) cation: K$^+$; anion: MnO$_4^-$, memorized; (+1) + (-1) = 0

(b) cation: K$^+$; anion: Cr$_2$O$_7^{2-}$, memorized; 2(+1) + (-2) = 0

(c) cation: Na$^+$; anion: CrO$_4^{2-}$, memorized; 2(+1) + (-2) = 0

(d) cation: Na$^+$; anion: ate/ic, chloric acid/HClO$_3$, ClO$_3^-$; (+1) + (-1) = 0

9.51 NH$_4$NO$_3$

cation: NH$_4^+$, memorized; anion: ate/ic, nitric acid/HNO$_3$, NO$_3^-$; (+1) + (-1) = 0

9.53 NaClO

cation: Na$^+$; anion: hypo-ite/hypo-ous, hypochlorous acid/HClO, ClO$^-$; (+1) + (-1) = 0

Chapter 10: Chemical Equations

10.1 (a) → reaction arrow meaning *changes, produces, yields*   (b) ⇌ reversible reaction

(c) ↑ product is a gas   (d) ↓ product drops out of solution as a precipitate or solid

(e) (aq) substance is dissolved in water   (f) (l) substance is a liquid   (g) (s) substance is a solid

(h) Δ heat is added   (i) (g) substance is gaseous

10.3 (a) $C_5H_{12} + 8\,O_2 \rightarrow 5\,CO_2 + 6\,H_2O$

(b) $Ca(OH)_2 + 2\,HCl \rightarrow CaCl_2 + 2\,H_2O$

(c) $CaCl_2 + 2\,AgNO_3 \rightarrow 2\,AgCl + Ca(NO_3)_2$

(d) $Na_2CO_3 + 2\,HBr \rightarrow 2\,NaBr + H_2O + CO_2$

(e) $4\,Al + 3\,O_2 \rightarrow 2\,Al_2O_3$

(a)   Skeletal equation: $C_5H_{12} + O_2 \rightarrow CO_2 + H_2O$

Step 1   Start with the most complicated formula, C_5H_{12}. (Balance H and O last.)

• Balance C: Starting with 5 C on the left and 1 C on the right, a total of 5 C are needed.

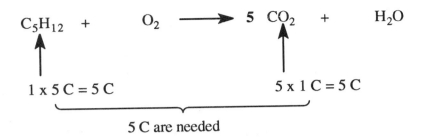

5 C are needed

Step 2

- Balance H: With 12 H on the left and 2 H on the right, a total of 12 H are needed.

  (Leave oxygen to the last, since it occurs as the element. Balancing O will not unbalance any other elements.)

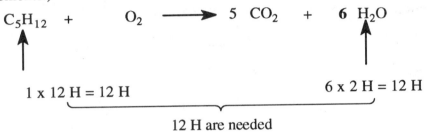

12 H are needed

- Balance O: With $10 + 6 = 16$ O on the right and 2 O on the left, 16 O are needed.

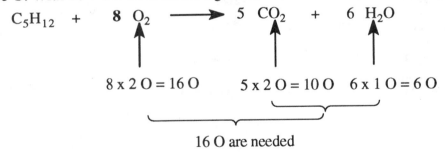

16 O are needed

(b) Skeletal equation: $Ca(OH)_2 + HCl \rightarrow CaCl_2 + H_2O$

Step 1 Start with $Ca(OH)_2$.

- Balance Ca: Ca is balanced.
- Balance O:

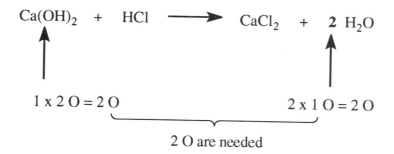

Step 2

- Balance Cl:

$$Ca(OH)_2 + \mathbf{2}\ HCl \longrightarrow CaCl_2 + 2\ H_2O$$

$$\underbrace{2 \times 1\ Cl = 2\ Cl \qquad 1 \times 2\ Cl = 2\ Cl}_{2\ Cl\ needed}$$

- Balance H: H is balanced.

(c) Skeletal equation: $CaCl_2 + AgNO_3 \rightarrow AgCl + Ca(NO_3)_2$

Step 1: Start with $Ca(NO_3)_2$.

- Balance Ca: Ca is balanced.

- Balance NO_3:

$$CaCl_2 + \mathbf{2}\ AgNO_3 \longrightarrow AgCl + Ca(NO_3)_2$$

$$\underbrace{2 \times 1\ NO_3 = 2\ NO_3 \qquad\qquad 2\ NO_3}_{2\ NO_3\ needed}$$

Step 2

- Balance Ag:

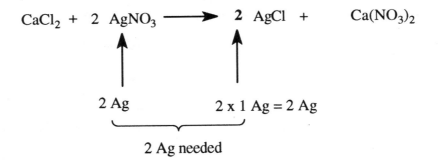

- Balance Cl: Cl is balanced.

(d) Skeletal equation: $Na_2CO_3 + HBr \rightarrow NaBr + H_2O + CO_2$

Step 1 Start with Na_2CO_3.

- Balance O: O is balanced.
- Balance C: C is balanced.
- Balance Na:

$$Na_2CO_3 + HBr \longrightarrow 2\ NaBr + H_2O + CO_2$$

2 Na   2 x 1 Na = 2 Na

2 Na needed

Step 2

- Balance Br:

$$Na_2CO_3 + 2\ HBr \longrightarrow 2\ NaBr + H_2O + CO_2$$

2 x 1 Br = 2 Br   2 Br

2 Br needed

- Balance H: H is balanced.

(e) Skeletal equation: $Al + O_2 \rightarrow Al_2O_3$

Step 1   Start with Al_2O_3.

- Balance Al:

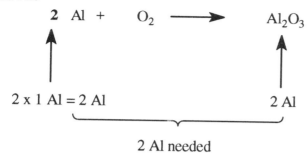

Step 2

- Balance O:

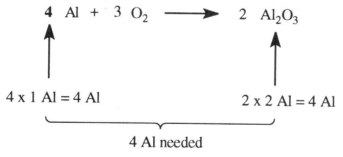

- Balance Al (no longer balanced):

10.5 (a) $4 NH_3 + 5 O_2 \rightarrow 4 NO + 6 H_2O$

(b) $Ni + 2 HCl \rightarrow NiCl_2 + H_2$

(c) $3 Ca(OH)_2 + 2 H_3PO_4 \rightarrow 6 H_2O + Ca_3(PO_4)_2$

(d) $2 NaI + Cl_2 \rightarrow 2 NaCl + I_2$

(a) Unbalanced equation: $NH_3 + O_2 \rightarrow NO + H_2O$

Step 1   Start with NH_3:

- Balance N: N is balanced.
- Balance H: With 3 H on the left and 2 H on the right, a total of 6 H is needed.

$$2\,NH_3 + O_2 \rightarrow NO + 3\,H_2O$$

- Balance N (no longer balanced):

$$2\,NH_3 + O_2 \rightarrow 2\,NO + 3\,H_2O$$

Step 2:

- Balance O: With 2 O on the left and 5 O on the right, a total of 10 O is needed. To obtain 10 O on the right, it will be necessary to double the coefficients of both products. Therefore it will also be necessary to double the coefficient of NH_3.

$$4\,NH_3 + 5\,O_2 \rightarrow 4\,NO + 6\,H_2O$$

(b) Unbalanced equation: $Ni + HCl \rightarrow NiCl_2 + H_2$

Step 1   Begin with $NiCl_2$.

- Balance Ni: Ni is balanced.
- Balance Cl: 2 Cl are needed on each side.

$$Ni + 2\,HCl \rightarrow NiCl_2 + H_2$$

Step 2

- Balance H: H is balanced.

(c) Unbalanced equation: $Ca(OH)_2 + H_3PO_4 \rightarrow H_2O + Ca_3(PO_4)_2$

Step 1   Begin with $Ca(PO_4)_3$.

- Balance Ca: 3 Ca are needed.

$$3\,Ca(OH)_2 + H_3PO_4 \rightarrow H_2O + Ca_3(PO_4)_2$$

- Balance PO_4 (actually PO_4^{3-} ion): 2 PO_4 are needed on each side.

$$3\,Ca(OH)_2 + 2\,H_3PO_4 \rightarrow H_2O + Ca_3(PO_4)_2$$

Step 2

- Balance H: 12 H are needed on each side.

$$3\ Ca(OH)_2\ +\ 2\ H_3PO_4\ \rightarrow\ \mathbf{6}\ H_2O\ +\ Ca_3(PO_4)_2$$

- Balance O: O is balanced.

(d) Unbalanced equation: $NaI\ +\ Cl_2\ \rightarrow\ NaCl\ +\ I_2$

Step 1   Begin with NaI

- Balance Na: Na is balanced.
- Balance I: 2 I are needed on each side.

$$\mathbf{2}\ NaI\ +\ Cl_2\ \rightarrow\ NaCl\ +\ I_2$$

- Balance Na (no longer balanced): 2 Na are needed on each side.

$$2\ NaI\ +\ Cl_2\ \rightarrow\ \mathbf{2}\ NaCl\ +\ I_2$$

Step 2

- Balance Cl: Cl is balanced.

10.7 (a) iron(II) acetate + potassium sulfide → iron(II) sulfide + potassium acetate

(b) dinitrogen tetroxide $\xrightarrow{\Delta}$ nitrogen dioxide

(c) mercury(I) nitrate + sodium chloride → mercury(I) chloride + sodium nitrate

10.9 (a) $Fe(C_2H_3O_2)_2(aq)\ +\ K_2S(aq)\ \rightarrow\ FeS(s)\ +\ 2\ KC_2H_3O_2(aq)$

(b) $N_2O_4(g)\ \xrightarrow{\Delta}\ 2\ NO_2(g)$

(c) $Hg_2(NO_3)_2(aq)\ +\ 2\ NaCl(aq)\ \rightarrow\ Hg_2Cl_2(s)\ +\ 2\ NaNO_3(aq)$

(a) Skeletal equation with correct formulas:

$$Fe(C_2H_3O_2)_2\ +\ K_2S\ \rightarrow\ FeS\ +\ KC_2H_3O_2$$

Step 1   Start with $Fe(C_2H_3O_2)_2$.

- Balance Fe: Fe is already balanced.
- Balance $C_2H_3O_2$: 2 $C_2H_3O_2$ are needed on each side.

$$Fe(C_2H_3O_2)_2 + K_2S \rightarrow FeS + 2\,KC_2H_3O_2$$

Step 2

- Balance K: K is balanced.
- Balance S: S is balanced.

Step 3   Add physical states.

$$Fe(C_2H_3O_2)_2(aq) + K_2S(aq) \rightarrow FeS(s) + 2\,KC_2H_3O_2(aq)$$

(b) Skeletal equation with correct formulas:

$$N_2O_4 \rightarrow NO_2$$

Step 1   Begin with N_2O_4.

- Balance N: 2 N are needed on each side.

$$N_2O_4 \rightarrow 2\,NO_2$$

- Balance O: O is balanced.

Step 2   Add physical states.

$$N_2O_4(g) \xrightarrow{\Delta} 2\,NO_2(g)$$

(c) Skeletal equation with correct formulas:

$$Hg_2(NO_3)_2 + NaCl \rightarrow Hg_2Cl_2 + NaNO_3$$

Step 1   Begin with $Hg_2(NO_3)_2$.

- Balance Hg: Hg is balanced.
- Balance NO_3: 2 NO_3 are needed on each side.

$$Hg_2(NO_3)_2 + NaCl \rightarrow Hg_2Cl_2 + 2\,NaNO_3$$

Step 2

- Balance Na: 2 Na are needed on each side.

$$Hg_2(NO_3)_2 + 2\,NaCl \rightarrow Hg_2Cl_2 + 2\,NaNO_3$$

- Balance Cl: Cl is balanced.

Step 3   Add physical states.

$$Hg_2(NO_3)_2(aq) + 2\, NaCl(aq) \rightarrow Hg_2Cl_2(s) + 2\, NaNO_3(aq)$$

10.11 (a)   sulfuric acid + iron → iron(II) sulfate + hydrogen

(b)   sodium chloride $\xrightarrow{\text{electrolysis}}$ sodium + chlorine

(c)   copper(I) sulfide + oxygen → copper + sulfur dioxide

10.13 $NaOCl(aq) + NaCl(aq) + 2\, HC_2H_3O_2(aq) \rightarrow Cl_2(g) + H_2O(l) + 2\, NaC_2H_3O_2(aq)$

1. Word equation:

sodium hypochlorite + sodium chloride + acetic acid → chlorine + water + sodium acetate

2. Skeletal equation with correct formulas:

$$NaOCl + NaCl + HC_2H_3O_2 \rightarrow Cl_2 + H_2O + NaC_2H_3O_2$$

3. Balanced equation:

Step 1   Begin with NaOCl.

- Balance Na: 2 Na are needed on each side.

$$NaOCl + NaCl + HC_2H_3O_2 \rightarrow Cl_2 + H_2O + \mathbf{2}\, NaC_2H_3O_2$$

- Balance O: O is balanced. (treat $C_2H_3O_2$ as a unit)

- Balance Cl: Cl is balanced.

Step 2

- Balance $C_2H_3O_2$: 2 $C_2H_3O_2$ are needed on each side.

$$NaOCl + NaCl + \mathbf{2}\, HC_2H_3O_2 \rightarrow Cl_2 + H_2O + 2\, NaC_2H_3O_2$$

- Balance H: H is balanced.

Step 3   Add physical states.

NaOCl(aq) + NaCl(aq) + 2 HC₂H₃O₂(aq) → Cl₂(g) + H₂O(l) + 2 NaC₂H₃O₂(aq)

10.15 (a) 3 KOH(aq) + H₃PO₄(aq) → 3 H₂O(l) + K₃PO₄(aq)

  (b) Mg(s) + NiCl₂(aq) → MgCl₂(aq) + Ni(s)

  (c) MgCO₃(s) + 2 HCl(aq) → H₂O(l) + MgCl₂(aq) + CO₂(g)

(a) 1. Word equation:

  potassium hydroxide + phosphoric acid → water + potassium phosphate

 2. Skeletal equation with correct formulas:

  KOH + H₃PO₄ → H₂O + K₃PO₄

 3. Balanced equation:

  Step 1   Begin with H₃PO₄.

  • Balance H: 4 H are needed on both sides.

  KOH + H₃PO₄ → 2 H₂O + K₃PO₄

  • Balance PO₄: PO₄ is balanced.

  Step 2

  • Balance K: 3 K are needed on each side.

  3 KOH + H₃PO₄ → 2 H₂O + K₃PO₄

  • Balance H (no longer balanced): 6 H are needed on each side.

  3 KOH + H₃PO₄ → 3 H₂O + K₃PO₄

  • Balance O: O is balanced.

  Step 3   Add physical states.

  3 KOH(aq) + H₃PO₄(aq) → 3 H₂O(l) + K₃PO₄(aq)

(b) 1. Word equation:

  magnesium + nickel chloride → magnesium chloride + nickel

Chapter 10: Chemical Equations

2. Skeletal equation with corrected formulas:

   $Mg + NiCl_2 \rightarrow MgCl_2 + Ni$

3. Balanced equation:

   Step 1   Begin with $NiCl_2$.

   - Balance Ni: Ni is balanced.
   - Balance Cl: Cl is balanced.

   Step 2

   - Balance Mg: Mg is balanced.

   Step 3   Add physical states.

   $Mg(s) + NiCl_2(aq) \rightarrow MgCl_2(aq) + Ni(s)$

(c) 1. Word equation:

magnesium carbonate + hydrochloric acid → water + magnesium chloride + carbon dioxide

2. Skeletal equation with correct formulas:

   $MgCO_3 + HCl \rightarrow H_2O + MgCl_2 + CO_2$

3. Balanced equation:

   Step 1   Begin with $MgCO_3$.

   - Balance Mg: Mg is balanced.
   - Balance C: C is balanced.
   - Balance O: O is balanced.

   Step 2

   - Balance H: 2 H are needed on each side.

     $MgCO_3 + \mathbf{2}\,HCl \rightarrow H_2O + MgCl_2 + CO_2$

   - Balance Cl: Cl is balanced.

   Step 3   Add physical states.

   $MgCO_3(s) + 2\,HCl(aq) \rightarrow H_2O(l) + MgCl_2(aq) + CO_2(g)$

10.17 (a) $2\,K(s) + 2\,H_2O(l) \rightarrow 2\,KOH(aq) + H_2(g)$

(b) $2\,Al(s) + 3\,H_2SO_4(aq) \rightarrow 3\,H_2(g) + Al_2(SO_4)_3(aq)$

(c) $MgCO_3(s) \rightarrow MgO(s) + CO_2(g)$

(d) $Na_2SO_4(aq) + BaCl_2(aq) \rightarrow 2\,NaCl(aq) + BaSO_4(s)$

(a) 1. Word equation:

potassium + water → potassium hydroxide + hydrogen

2. Skeletal equation with correct formulas:

$K + H_2O \rightarrow KOH + H_2$

3. Balanced equation:

Step 1 Begin with KOH.

- Balance K: K is balanced.
- Balance O: O is balanced.
- Balance H: With 2 H on the left and 3 H on the right, it looks as if 6 H are needed on both sides. However, since one of the products on the right is the element H, the equation can temporarily be balanced by using the fraction $\frac{1}{2}$ for H_2.

$K + H_2O \rightarrow KOH + \frac{1}{2}H_2$

Step 2 To remove fractions, multiply the entire equation by a factor of 2.

$2\,K + 2\,H_2O \rightarrow 2\,KOH + H_2$

Step 3 Add physical states.

$2\,K(s) + 2\,H_2O(l) \rightarrow 2\,KOH(aq) + H_2(g)$

(b) 1. Word equation:

aluminum + sulfuric acid → hydrogen + aluminum sulfate

2. Skeletal equation with correct formulas:

$$Al + H_2SO_4 \rightarrow H_2 + Al_2(SO_4)_3$$

3. Balanced equation:

   Step 1   Begin with $Al_2(SO_4)_3$.

   - Balance Al: 2 Al are needed on each side.

   $$\mathbf{2}\,Al + H_2SO_4 \rightarrow H_2 + Al_2(SO_4)_3$$

   - Balance SO_4: 3 SO_4 are needed on each side.

   $$2\,Al + \mathbf{3}\,H_2SO_4 \rightarrow H_2 + Al_2(SO_4)_3$$

   Step 2

   - Balance H: 6 H are needed on each side.

   $$2\,Al + 3\,H_2SO_4 \rightarrow \mathbf{3}\,H_2 + Al_2(SO_4)_3$$

   Step 3   Add physical states.

   $$2\,Al(s) + 3\,H_2SO_4(aq) \rightarrow 3\,H_2(g) + Al_2(SO_4)_3(aq)$$

(c) 1. Word equation:

   magnesium carbonate → magnesium oxide + carbon dioxide

   2. Skeletal equation with correct formulas:

   $$MgCO_3 \rightarrow MgO + CO_2$$

   3. Balanced equation:

   Step 1   Start with $MgCO_3$.

   - Balance Mg: Mg is balanced.
   - Balance C: C is balanced.
   - Balance O: O is balanced.

   Step 2   Add physical states.

   $$MgCO_3(s) \rightarrow MgO(s) + CO_2(g)$$

(d) 1. Word equation:

   sodium sulfate + barium chloride → sodium chloride + barium sulfate

2. Skeletal equation with correct formulas:

$$Na_2SO_4 + BaCl_2 \rightarrow NaCl + BaSO_4$$

3. Balanced equation:

Step 1   Begin with Na_2SO_4.

- Balance Na: 2 Na are needed on each side.

$$Na_2SO_4 + BaCl_2 \rightarrow \mathbf{2}\ NaCl + BaSO_4$$

- Balance SO_4: SO_4 is balanced.

Step 2

- Balance Ba: Ba is balanced.
- Balance Cl: Cl is balanced.

Step 3   Add physical states.

$$Na_2SO_4(aq) + BaCl_2(aq) \rightarrow 2\ NaCl(aq) + BaSO_4(s)$$

10.19 (a) $N_2(g) + O_2(g) \rightarrow 2\ NO(g)$

(b) $SO_3(g) + H_2O(l) \rightarrow H_2SO_4(aq)$

(c) $P_4(s) + 6\ Cl_2(g) \rightarrow 4\ PCl_3(s)$

10.21 (a) $2\ Na_2O_2(s) \rightarrow 2\ Na_2O(s) + O_2(g)$

(b) $H_2CO_3(aq) \rightarrow CO_2(g) + H_2O(l)$

10.23 (a) $C_6H_{12}O_6(s) + 6\ O_2(g) \rightarrow 6\ CO_2(g) + 6\ H_2O(g)$

(b) $2\ C_8H_{18}(l) + 25\ O_2(g) \rightarrow 16\ CO_2(g) + 18\ H_2O(g)$

(a) 1.   Word equation:

glucose + oxygen → carbon dioxide + water

2. Correct formulas:

$$C_6H_{12}O_6 + O_2 \rightarrow CO_2 + H_2O$$

3. Balanced equation:

Step 1   Start with $C_6H_{12}O_6$.

- Balance C and H: 6 C and 12 H are needed.

$$C_6H_{12}O_6 + O_2 \rightarrow \mathbf{6}\,CO_2 + \mathbf{6}\,H_2O$$

Step 2

- Balance O: 18 O are needed on each side. Don't overlook the O's in the glucose formula.

$$C_6H_{12}O_6 + \mathbf{6}\,O_2 \rightarrow 6\,CO_2 + 6\,H_2O$$

Step 3   Add physical states.

$$C_6H_{12}O_6(s) + 6\,O_2(g) \rightarrow 6\,CO_2(g) + 6\,H_2O(g)$$

(b) 1. Word equation:

octane + oxygen → carbon dioxide + water

2. Correct formulas:

$$C_8H_{18} + O_2 \rightarrow CO_2 + H_2O$$

3. Balanced equation:

Step 1   Start with C_8H_{18}.

- Balance C and H: 8 C and 18 H are needed.

$$C_8H_{18} + O_2 \rightarrow \mathbf{8}\,CO_2 + \mathbf{9}\,H_2O$$

Step 2

- Balance O: 25 O are needed on each side.

$$C_8H_{18} + \tfrac{25}{2}\,O_2 \rightarrow 8\,CO_2 + 9\,H_2O$$

Step 3   To clear the fraction, multiply all coefficients by 2. Add physical states.

$$2\,C_8H_{18}(l) + 25\,O_2(g) \rightarrow 16\,CO_2(g) + 18\,H_2O(g)$$

10.25 $C_7H_8(l) + 9\ O_2(g) \rightarrow 7\ CO_2(g) + 4\ H_2O(g)$

1. Word equation:

    toluene + oxygen → carbon dioxide + water

2. Correct formulas:

    $$C_7H_8 + O_2 \rightarrow CO_2 + H_2O$$

3. Balanced equation:

    Step 1   Start with C_7H_8.

    - Balance C and H: 7 C and 8 H are needed on each side.

    $$C_7H_8 + O_2 \rightarrow 7\ CO_2 + 4\ H_2O$$

    Step 2

    - Balance O: 18 O are needed on each side.

    $$C_7H_8 + 9\ O_2 \rightarrow 7\ CO_2 + 4\ H_2O$$

    Step 3   Add physical states.

    $$C_7H_8(l) + 9\ O_2(g) \rightarrow 7\ CO_2(g) + 4\ H_2O(g)$$

10.27 (a) $Zn(s) + H_2SO_4(aq) \rightarrow ZnSO_4(aq) + H_2(g)$ (b) $Ca(s) + 2\ H_2O(l) \rightarrow Ca(OH)_2(aq) + H_2(g)$

(c) $Al(s) + H_2O(l) \rightarrow$ no reaction   (d) $Mg(s) + 2\ AgNO_3(aq) \rightarrow Mg(NO_3)_2(aq) + 2\ Ag(s)$

(a) Zinc will displace hydrogen from acids.

$$Zn(s) + H_2SO_4(aq) \rightarrow ZnSO_4(aq) + H_2(g)$$

(b) Calcium will displace hydrogen from water.

$$Ca(s) + 2\ H_2O(l) \rightarrow Ca(OH)_2(aq) + H_2(g)$$

(c) Aluminum is below hydrogen in the activity series, so no reaction occurs.

(d) Magnesium is above silver in the activity series, so it will displace silver from its compounds.

Mg(s) + 2 AgNO$_3$(aq) → Mg(NO$_3$)$_2$(aq) + 2 Ag(s)

10.29 (a) AgNO$_3$(aq) + HCl(aq) → AgCl(s) + HNO$_3$(aq)
(b) CuCl$_2$(aq) + Na$_2$S(aq) → CuS(s) + 2 NaCl(aq)
(c) Fe(NO$_3$)$_3$(aq) + NaCl(aq) → no reaction
(d) Pb(NO$_3$)$_2$(aq) + K$_2$CrO$_4$(aq) → PbCrO$_4$(s) + 2 KNO$_3$(aq)

(a) Silver chloride is insoluble. Therefore a reaction will occur.

AgNO$_3$(aq) + HCl(aq) → AgCl(s) + HNO$_3$(aq)

(b) Copper(II) sulfide is insoluble. Therefore a reaction will occur.

CuCl$_2$(aq) + Na$_2$S(aq) → CuS(s) + 2 NaCl(aq)

(c) Since neither iron(III) chloride (FeCl$_3$) nor sodium nitrate (NaNO$_3$) is insoluble, no reaction will occur.

(d) Since lead(II) chromate is insoluble, a reaction will occur.

Pb(NO$_3$)$_2$(aq) + K$_2$CrO$_4$(aq) → PbCrO$_4$(s) + 2 KNO$_3$(aq)

10.31 (a) 2 Na$_3$PO$_4$(aq) + 3 CaCl$_2$(aq) → Ca$_3$(PO$_4$)$_2$(s) + 6 NaCl(aq)
(b) K$_2$SO$_4$(aq) + Ba(NO$_3$)$_2$(aq) → BaSO$_4$(s) + 2 KNO$_3$(aq)
(c) Hg(NO$_3$)$_2$(aq) + NaCl(aq) → no reaction
(d) 2 Al(NO$_3$)$_3$(aq) + 3 Na$_2$S(aq) → Al$_2$S$_3$(s) + 6 NaNO$_3$(aq)

(a) Since calcium phosphate is insoluble, a precipitation reaction will occur.

2 Na$_3$PO$_4$(aq) + 3 CaCl$_2$(aq) → Ca$_3$(PO$_4$)$_2$(s) + 6 NaCl(aq)

(b) Since barium sulfate is insoluble, a precipitation reaction will occur.

$$K_2SO_4(aq) + Ba(NO_3)_2(aq) \rightarrow BaSO_4(s) + 2\ KNO_3(aq)$$

(c) Since neither mercury(II) chloride ($HgCl_2$) nor sodium nitrate ($NaNO_3$) is insoluble, no reaction occurs.

(d) Since aluminum sulfide is insoluble, a precipitation reaction will occur.

$$2\ Al(NO_3)_3(aq) + 3\ Na_2S(aq) \rightarrow Al_2S_3(s) + 6\ NaNO_3(aq)$$

10.33 (a) $Ag^+(aq) + Cl^-(aq) \rightarrow AgCl(s)$ (b) $Cu^{2+}(aq) + S^{2-}(aq) \rightarrow CuS(s)$

(c) No net reaction. (d) $Pb^{2+}(aq) + CrO_4^{2-}(aq) \rightarrow PbCrO_4(s)$

(a) *Total equation:*

$$AgNO_3(aq) + HCl(aq) \rightarrow AgCl(s) + HNO_3(aq)$$

Total ionic equation:

$$Ag^+(aq) + NO_3^-(aq) + H^+(aq) + Cl^-(aq) \rightarrow AgCl(s) + H^+(aq) + NO_3^-(aq)$$

Net ionic equation:

$$Ag^+(aq) + Cl^-(aq) \rightarrow AgCl(s)$$

(b) *Total equation:*

$$CuCl_2(aq) + Na_2S(aq) \rightarrow CuS(s) + 2\ NaCl(aq)$$

Total ionic equation:

$$Cu^{2+}(aq) + 2\ Cl^-(aq) + 2\ Na^+(aq) + S^{2-}(aq) \rightarrow CuS(s) + 2\ Na^+(aq) + 2\ Cl^-(aq)$$

Net ionic equation:

$$Cu^{2+}(aq) + S^{2-}(aq) \rightarrow CuS(s)$$

(c) *Total equation:*

$$Fe(NO_3)_3(aq) + 3\ NaCl(aq) \rightarrow FeCl_3(aq) + 3\ NaNO_3(aq)$$

Total ionic equation:

$$Fe^{3+}(aq) + 3\ NO_3^-(aq) + 3\ Na^+(aq) + 3\ Cl^-(aq) \rightarrow Fe^{3+}(aq) + 3\ Cl^-(aq) + 3\ Na^+(aq) + 3\ NO_3^-(aq)$$

Net ionic equation:

no net reaction

(d) *Total equation:*

Pb(NO$_3$)$_2$(aq) + K$_2$CrO$_4$(aq) → PbCrO$_4$(s) + 2 KNO$_3$(aq)

Total ionic equation:

Pb^{2+}(aq) + 2 NO$_3^-$(aq) + 2 K$^+$(aq) + CrO$_4^{2-}$(aq) → PbCrO$_4$(s) + 2 K$^+$(aq) + 2 NO$_3^-$(aq)

Net ionic equation:

Pb^{2+}(aq) + CrO$_4^{2-}$(aq) → PbCrO$_4$(s)

10.35 2 Na(s) + 2 H$_2$O(l) → 2 Na$^+$(aq) + 2 OH$^-$(aq) + H$_2$(g)

Total equation:

2 Na(s) + 2 H$_2$O(l) → 2 NaOH(aq) + H$_2$(g)

Total ionic equation:

2 Na(s) + 2 H$_2$O(l) → 2 Na$^+$(aq) + 2 OH$^-$(aq) + H$_2$(g)

Net ionic equation:

2 Na(s) + 2 H$_2$O(l) → 2 Na$^+$(aq) + 2 OH$^-$(aq) + H$_2$(g) [identical to the total ionic equation]

10.37 NH$_4^+$(aq) + OH$^-$(aq) → NH$_3$(g) + H$_2$O(l)

Total equation:

NH$_4$Cl(aq) + NaOH(aq) → NaCl(aq) + NH$_3$(g) + H$_2$O(l)

Total ionic equation:

NH$_4^+$(aq) + Cl$^-$(aq) + Na$^+$(aq) + OH$^-$(aq) → Na$^+$(aq) + Cl$^-$(aq) + NH$_3$(g) + H$_2$O(l)

Net ionic equation:

$NH_4^+(aq) + OH^-(aq) \rightarrow NH_3(g) + H_2O(l)$

10.39 (a) $2\ Ca(s) + O_2(g) \rightarrow 2\ CaO(s)$

(b) $2\ HClO_3(aq) + Ba(OH)_2(aq) \rightarrow Ba(ClO_3)_2(aq) + 2\ H_2O(l)$

(c) $2\ HgO(s) \rightarrow O_2(g) + 2\ Hg(l)$

(d) $Ca(s) + 2\ H_2O(l) \rightarrow H_2(g) + Ca(OH)_2(s)$

(a) This is a combination reaction. The formula of the product is CaO because calcium forms a 2+ ion and oxygen forms a 2- ion.

$2\ Ca(s) + O_2(g) \rightarrow 2\ CaO(s)$

(b) This is a double replacement, neutralization reaction.

$2\ HClO_3(aq) + Ba(OH)_2(aq) \rightarrow Ba(ClO_3)_2(aq) + 2\ H_2O(l)$

(c) This is a decomposition reaction. The other product must be mercury.

$2\ HgO(s) \rightarrow O_2(g) + 2\ Hg(l)$

(d) This is a single replacement reaction. Calcium is more reactive than the hydrogen in water.

$Ca(s) + 2\ H_2O(l) \rightarrow H_2(g) + Ca(OH)_2(s)$

10.41 (a) combustion (b) double replacement (c) double replacement

(d) double replacement/decomposition (e) combination

(a) This reaction is recognized as a combustion reaction by the presence of O_2 as a reactant and the identity of the products, CO_2 and H_2O. (b, c) These reactions are recognized as double replacement by the exchange of anions between two cations. (d) This reaction is not immediately obvious as to classification. It is clear that the sodium ion has exchanged anions. Since single replacement is ruled out by the presence of two compounds as reactants, double replacement is a logical assumption. If double replacement is assumed, then the other product besides NaBr would have been H_2CO_3. Since H_2CO_3 is known to decompose to CO_2 and H_2O, the reaction is classified as double replacement followed by decomposition. (e) When two substances react to form one compound, the reaction is classified as combination.

10.43 (a) $Mg(s) + 2 HCl(aq) \rightarrow MgCl_2(aq) + H_2(g)$

(b) $2 C_6H_6(l) + 15 O_2(g) \rightarrow 12 CO_2(g) + 6 H_2O(g)$

(c) $Cd(NO_3)_2(aq) + H_2S(aq) \rightarrow CdS(s) + 2 HNO_3(aq)$

(d) $FeCl_3(aq) + Na_3PO_4(aq) \rightarrow FePO_4(s) + 3 NaCl(aq)$

(a) This is a single replacement reaction. Magnesium displaces hydrogen.

$Mg(s) + 2 HCl(aq) \rightarrow MgCl_2(aq) + H_2(g)$

(b) This is a combustion reaction. The products are carbon dioxide and water.

$2 C_6H_6(l) + 15 O_2(g) \rightarrow 12 CO_2(g) + 6 H_2O(l)$

(c) This is a double replacement, precipitation reaction.

$Cd(NO_3)_2(aq) + H_2S(aq) \rightarrow CdS(s) + 2 HNO_3(aq)$

(d) This is a double replacement, precipitation reaction.

$FeCl_3(aq) + Na_3PO_4(aq) \rightarrow FePO_4(s) + 3 NaCl(aq)$

10.45 (a) oxidized: Mg; reduced: H in HCl (b) oxidized: C in C_6H_6; reduced: O

(a) Mg is oxidized, since its oxidation number increased from 0 to +2. H is reduced, since its oxidation number decreased from +1 to 0. (b) The oxidation number of the C in C_6H_6 is -1 each. The oxidation number of C in CO_2 is +4. Therefore the carbon is oxidized. The oxidation number of the oxygen decreases from 0 to -2, which is reduction.

10.47 (a) $Al(NO_3)_3(aq) + 3\,NaOH(aq) \rightarrow Al(OH)_3(s) + 3\,NaNO_3(aq)$

(b) $Ni(s) + 2\,HBr(aq) \rightarrow NiBr_2(aq) + H_2(g)$

(c) $C_4H_{10}O(l) + 6\,O_2(g) \rightarrow 4\,CO_2(g) + 5\,H_2O(g)$

(a) This is a double replacement, precipitation reaction.

$Al(NO_3)_3(aq) + 3\,NaOH(aq) \rightarrow Al(OH)_3(s) + 3\,NaNO_3(aq)$

(b) This is a single replacement reaction. Nickel is above hydrogen in the activity series.

$Ni(s) + 2\,HBr(aq) \rightarrow NiBr_2(aq) + H_2(g)$

(c) This is a combustion reaction. The products are carbon dioxide and water.

$C_4H_{10}O(l) + 6\,O_2(g) \rightarrow 4\,CO_2(g) + 5\,H_2O(g)$

10.49 Nothing happened to the ring because gold is below hydrogen in the activity series. The ring could be recovered by carefully pouring off the acid into another beaker, leaving the ring in the original beaker. Then the ring should be thoroughly rinsed with water before handling.

Chapter 11:
Calculations Involving Chemical Equations

11.1 6 mol H_2

Find: number of moles H_2 = ?

Given: 3 mol O_2

Known: $\dfrac{2 \text{ mol } H_2}{1 \text{ mol } O_2}$ from balanced equation

Solution: mol H_2 = $(\dfrac{2 \text{ mol } H_2}{1 \text{ mol } O_2})(3 \text{ mol } O_2)$ = 6 mol H_2

11.3 3 mol O_2

Find: number of moles of O_2 = ?

Given: 6 mol H_2O_2

Known: products of decomposition are water and oxygen

balanced equation: $2 H_2O_2 \rightarrow 2 H_2O + O_2$

mole ratio: $\dfrac{1 \text{ mol } O_2}{2 \text{ mol } H_2O_2}$

Solution: mol O_2 = $(\dfrac{1 \text{ mol } O_2}{2 \text{ mol } H_2O_2})(6 \text{ mol } H_2O_2)$ = 3 mol O_2

110   Student's Solutions Manual

11.5   2.6 mol NaHCO$_3$

Find: number of moles of NaHCO$_3$ = ?

Given: 1.3 mol H$_2$SO$_4$

Known: $\dfrac{2 \text{ mol NaHCO}_3}{1 \text{ mol H}_2\text{SO}_4}$ from balanced equation

Solution: moles NaHCO$_3$ = $(\dfrac{2 \text{ mol NaHCO}_3}{1 \text{ mol H}_2\text{SO}_4})(1.3 \text{ mol H}_2\text{SO}_4)$ = 2.6 mol NaHCO$_3$

11.7   (a) 3.28 mol O$_2$ (b) 1.50 mol P$_4$O$_{10}$

(a) Find: number of moles of O$_2$ = ?

Given: 0.655 mol P$_4$

Known: balanced equation: P$_4$(s) + 5 O$_2$(g) → P$_4$O$_{10}$(s)

   mole ratio: $\dfrac{5 \text{ mol O}_2}{1 \text{ mol P}_4}$

Solution: mol O$_2$ = $(\dfrac{5 \text{ mol O}_2}{1 \text{ mol P}_4})(0.655 \text{ mol P}_4)$ = 3.28 mol O$_2$

(b) Find: number of moles of P$_4$O$_{10}$ = ?

Given: 1.50 mol P$_4$

Known: $\dfrac{1 \text{ mol P}_4\text{O}_{10}}{1 \text{ mol P}_4}$ from balanced equation above

Solution: mol P$_4$O$_{10}$ = $(\dfrac{1 \text{ mol P}_4\text{O}_{10}}{1 \text{ mol P}_4})(1.50 \text{ mol P}_4)$ = 1.50 mol P$_4$O$_{10}$

11.9   (a) 0.250 mol Na$_2$SO$_3$ (b) 0.500 mol HCl

(a) Find: number of moles of Na$_2$SO$_3$ = ?

Given: 0.250 mol Cl$_2$

Known: $\dfrac{1 \text{ mol Na}_2\text{SO}_3}{1 \text{ mol Cl}_2}$ from balanced equation

Solution: mol Na$_2$SO$_3$ = $(\dfrac{1 \text{ mol Na}_2\text{SO}_3}{1 \text{ mol Cl}_2})(0.250 \text{ mol Cl}_2)$ = 0.250 mol Na$_2$SO$_3$

(b) Find: number of moles of HCl = ?

Given: 0.250 mol Cl_2

Known: $\dfrac{2 \text{ mol HCl}}{1 \text{ mol Cl}_2}$ from balanced equation

Solution: mol HCl = $(\dfrac{2 \text{ mol HCl}}{1 \text{ mol Cl}_2})(0.250 \text{ mol Cl}_2)$ = 0.500 mol HCl

11.11 (a) 12.7 mol C_2H_5OH (b) 12.7 mol CO_2

(a) Find: number of moles of C_2H_5OH = ?

Given: 6.35 mol $C_6H_{12}O_6$

Known: $\dfrac{2 \text{ mol C}_2\text{H}_5\text{OH}}{1 \text{ mol C}_6\text{H}_{12}\text{O}_6}$ from balanced equation

Solution: mol C_2H_5OH = $(\dfrac{2 \text{ mol C}_2\text{H}_5\text{OH}}{1 \text{ mol C}_6\text{H}_{12}\text{O}_6})(6.35 \text{ mol C}_6\text{H}_{12}\text{O}_6)$ = 12.7 mol C_2H_5OH

(b) Find: number of moles of CO_2 = ?

Given: 6.35 mol $C_6H_{12}O_6$

Known: $\dfrac{2 \text{ mol CO}_2}{1 \text{ mol C}_6\text{H}_{12}\text{O}_6}$ from balanced equation

Solution: mol CO_2 = $(\dfrac{2 \text{ mol CO}_2}{1 \text{ mol C}_6\text{H}_{12}\text{O}_6})(6.35 \text{ mol C}_6\text{H}_{12}\text{O}_6)$ = 12.7 mol CO_2

11.13   928 g CO_2

First write the word equation. Convert it to a balanced chemical equation. Outline the problem under the balanced equation. Calculate molar masses from atomic masses.

methanol   + oxygen   →   carbon dioxide   + water

$CH_3OH(l) \quad + \quad 2\,O_2(g) \quad \rightarrow \quad CO_2(g) \quad + \quad 2\,H_2O(l)$

675 g   ? g

MM = 32.0 g/mol   MM = 44.0 g/mol

_____ mol $CH_3OH \times \dfrac{1 \text{ mol } CO_2}{1 \text{ mol } CH_3OH} \rightarrow$ _____ mol CO_2

This problem involves three equations:

number of moles $CH_3OH = (\dfrac{1 \text{ mol } CH_3OH}{32.0 \text{ g } CH_3OH})(675 \text{ g } CH_3OH) = 21.1$ mol CH_3OH

number of moles $CO_2 = (\dfrac{1 \text{ mol } CO_2}{1 \text{ mol } CH_3OH})(21.1 \text{ mol } CH_3OH) = 21.1$ mol CO_2

mass (g) $CO_2 = (\dfrac{44.0 \text{ g } CO_2}{1 \text{ mol } CO_2})(21.1 \text{ mol } CO_2) = 928$ g CO_2

11.15 6.30×10^2 g $NaHCO_3$

Outline the problem under the balanced equation. Calculate molar masses from atomic masses.

$H_2SO_4(aq) \quad + \quad 2\,NaHCO_3(s) \quad \rightarrow \quad Na_2SO_4(aq) \quad + 2\,CO_2(g) \quad + \quad 2\,H_2O(l)$

? g

MM = 84.0 g/mol

_____ mol $NaHCO_3 \leftarrow \dfrac{2 \text{ mol } NaHCO_3}{2 \text{ mol } CO_2} \times 7.50$ mol CO_2

This problem involves two calculations:

number of moles $NaHCO_3 = (\dfrac{2 \text{ mol } NaHCO_3}{2 \text{ mol } CO_2})(7.50 \text{ mol } CO_2) = 7.50$ mol $NaHCO_3$

mass (g) $NaHCO_3 = (\dfrac{84.0 \text{ g } NaHCO_3}{1 \text{ mol } NaHCO_3})(7.50 \text{ mol } NaHCO_3) = 630.$ g $NaHCO_3$

11.17   0.405 g AgCl

First balance the equation. Outline the problem under the balanced equation. Calculate molar masses from atomic masses.

2 AgNO$_3$(aq) + CaCl$_2$(aq) → 2 AgCl(s) + Ca(NO$_3$)$_2$(aq)

0.478 g   ? g

MM = 170. g/mol   MM = 144 g/mol

____ mol AgNO$_3$ × $\dfrac{2 \text{ mol AgCl}}{2 \text{ mol AgNO}_3}$ → ____ mol AgCl

Three calculations are needed to solve this problem.

number of moles AgNO$_3$ = $(\dfrac{1 \text{ mol AgNO}_3}{170. \text{ g AgNO}_3})(0.478 \text{ g AgNO}_3)$ = 2.81 × 10^{-3} mol AgNO$_3$

number of moles AgCl = $(\dfrac{2 \text{ mol AgCl}}{2 \text{ mol AgNO}_3})(2.81 \times 10^{-3} \text{ mol AgNO}_3)$ = 2.81 × 10^{-3} mol AgCl

mass (g) AgCl = $(\dfrac{144 \text{ g AgCl}}{1 \text{ mol AgCl}})(2.81 \times 10^{-3} \text{ mol AgCl})$ = 0.405 g AgCl

11.19 0.22 g MgO

First write the word equation. Convert it to a balanced chemical equation. Outline the problem under the balanced equation. Calculate molar masses from atomic masses.

2 Mg(s) + O$_2$(g) → 2 MgO(s)

0.13 g   ? g

MM = 24 g/mol   MM = 40. g/mol

____ mol Mg × $\dfrac{2 \text{ mol MgO}}{2 \text{ mol Mg}}$ → ____ mol MgO

The three calculations may be combined into one:

mass (g) MgO = $(\frac{40.\text{ g MgO}}{1\text{ mol MgO}})(\frac{2\text{ mol MgO}}{2\text{ mol Mg}})(\frac{1\text{ mol Mg}}{24\text{ g Mg}})(0.13\text{ g Mg}) = 0.22$ g MgO

11.21 21.9 g CuO

Outline the problem under the balanced equation. Calculate molar masses from atomic masses.

CuO(s) + H$_2$(g) → H$_2$O(g) + Cu(s)

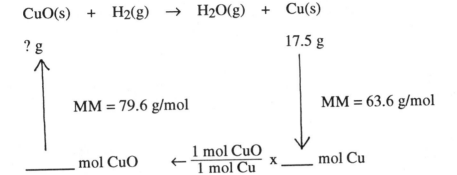

Combine the three calculations:

mass (g) CuO = $(\frac{79.6\text{ g CuO}}{1\text{ mol CuO}})(\frac{1\text{ mol CuO}}{1\text{ mol Cu}})(\frac{1\text{ mol Cu}}{63.6\text{ g Cu}})(17.5\text{ g Cu}) = 21.9$ g CuO

11.23 8.42 kg CaO

First write the balanced chemical equation from the word equation. Outline the problem under the balanced equation. Calculate molar masses from atomic masses. Note that 15.0 kg is equivalent to 15.0 x 10^3 g.

calcium carbonate → carbon dioxide + calcium oxide

$CaCO_3(s) \rightarrow CO_2(g) + CaO(s)$

15.0 x 10³ g   ? kg

MM = 100. g/mol   MM = 56.1 g/mol

_____ mol $CaCO_3$ × $\dfrac{1 \text{ mol CaO}}{1 \text{ mol } CaCO_3}$ → _____ mol CaO

Combine the three calculations:

mass (g) CaO = $(\dfrac{56.1 \text{ g CaO}}{1 \text{ mol CaO}})(\dfrac{1 \text{ mol CaO}}{1 \text{ mol } CaCO_3})(\dfrac{1 \text{ mol } CaCO_3}{100. \text{ g } CaCO_3})(15.0 \times 10^3 \text{ g } CaCO_3)$

= 8.42 × 10³ g CaO = 8.42 kg CaO

11.25 1.27 × 10³ g P_4

Outline the problem under the balanced equation. Calculate molar masses from atomic masses.

$P_4(s) + 6 Cl_2(g) \rightarrow 4 PCl_3(s)$

? g   5.67 × 10³ g

MM = 124 g/mol   MM = 138 g/mol

_____ mol P_4 ← $\dfrac{1 \text{ mol } P_4}{4 \text{ mol } PCl_3}$ × _____ mol PCl_3

mass (g) P_4 = $(\dfrac{124 \text{ g } P_4}{1 \text{ mol } P_4})(\dfrac{1 \text{ mol } P_4}{4 \text{ mol } PCl_3})(\dfrac{1 \text{ mol } PCl_3}{138 \text{ g } PCl_3})(5.67 \times 10^3 \text{ g } PCl_3)$ = 1.27 × 10³ g P_4

11.27 88.5 g Na and 137 g Cl_2

Outline the problem under the balanced equation. Calculate molar masses from atomic masses. Solve for one of the two elements first. The mass of the second element can be obtained by subtracting the mass of the other element from the mass of the sodium chloride.

$$2 \text{ NaCl(l)} \rightarrow \text{Cl}_2(g) + 2 \text{ Na(l)}$$

225 g   ? g

MM = 58.5 g/mol   MM = 23.0 g/mol

_____ mol NaCl × $\dfrac{2 \text{ mol Na}}{2 \text{ mol NaCl}}$ → _____ mol Na

mass (g) Na = $(\dfrac{23.0 \text{ g Na}}{1 \text{ mol Na}})(\dfrac{2 \text{ mol Na}}{2 \text{ mol NaCl}})(\dfrac{1 \text{ mol NaCl}}{58.5 \text{ g NaCl}})(225 \text{ g NaCl})$ = 88.5 g Na

mass (g) Cl$_2$ = mass NaCl - mass Na = 225 g - 88.5 g = 137 g Cl$_2$

11.29 43.7 g Fe$_2$O$_3$

Outline the problem under the balanced equation. Calculate molar masses from atomic masses.

$$4 \text{ FeS}_2(s) + 11 \text{ O}_2(g) \rightarrow 2 \text{ Fe}_2\text{O}_3(s) + 8 \text{ SO}_2(g)$$

65.6 g   ? g

MM = 120. g/mol   MM = 160. g/mol

_____ mol FeS$_2$ × $\dfrac{2 \text{ mol Fe}_2\text{O}_3}{4 \text{ mol FeS}_2}$ → _____ mol Fe$_2$O$_3$

mass (g) Fe$_2$O$_3$ = $(\dfrac{160. \text{ g Fe}_2\text{O}_3}{1 \text{ mol Fe}_2\text{O}_3})(\dfrac{2 \text{ mol Fe}_2\text{O}_3}{4 \text{ mol FeS}_2})(\dfrac{1 \text{ mol FeS}_2}{120. \text{ g FeS}_2})(65.6 \text{ g FeS}_2)$ = 43.7 g Fe$_2$O$_3$

11.31 4.19 g Ag

Outline the problem under the balanced equation. Calculate molar masses from atomic masses.

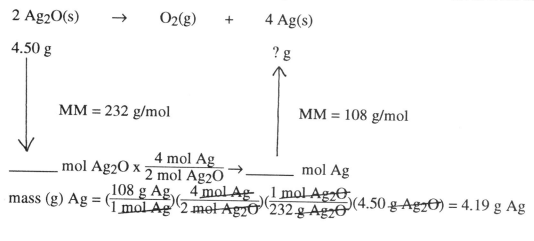

2 Ag₂O(s)   →   O₂(g)   +   4 Ag(s)
4.50 g   ? g

MM = 232 g/mol   MM = 108 g/mol

_____ mol Ag₂O × $\frac{4 \text{ mol Ag}}{2 \text{ mol Ag}_2\text{O}}$ → _____ mol Ag

mass (g) Ag = ($\frac{108 \text{ g Ag}}{1 \text{ mol Ag}}$)($\frac{4 \text{ mol Ag}}{2 \text{ mol Ag}_2\text{O}}$)($\frac{1 \text{ mol Ag}_2\text{O}}{232 \text{ g Ag}_2\text{O}}$)(4.50 g Ag₂O) = 4.19 g Ag

11.33 485 g CCl₄

Outline the problem under the balanced equation. Calculate molar masses from atomic masses.

CH₄(g)   +   4 Cl₂(g)   →   CCl₄(l)   +   4 HCl(g)
50.4 g   ? g

MM = 16.0 g/mol   MM = 154 g/mol

_____ mol CH₄ × $\frac{1 \text{ mol CCl}_4}{1 \text{ mol CH}_4}$ → _____ mol CCl₄

mass (g) CCl₄ = ($\frac{154 \text{ g CCl}_4}{1 \text{ mol CCl}_4}$)($\frac{1 \text{ mol CCl}_4}{1 \text{ mol CH}_4}$)($\frac{1 \text{ mol CH}_4}{16.0 \text{ g CH}_4}$)(50.4 g CH₄) = 485 g CCl₄

11.35 theoretical yield: 1.34 g Hg; percent yield: 92.5%

First determine the theoretical yield in the normal fashion by outlining the problem under the balanced equation. Calculate molar masses from atomic masses.

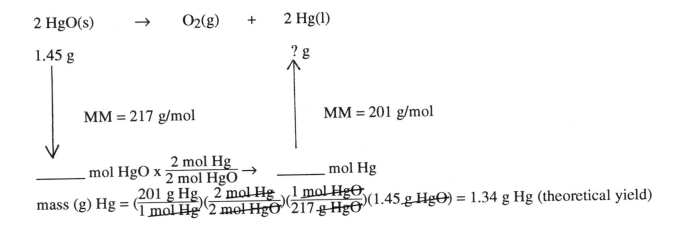

Find: percentage yield (%) = ?

Given: actual yield = 1.24 g Hg

Known: theoretical yield = 1.34 g Hg
Solution: $(\dfrac{1.24 \text{ g actual}}{1.34 \text{ g theoretical}})(100) = 92.5\%$

11.37 (a) 95.8 g Ti   (b) 74.1%

(a) First write the word equation. Convert it to a balanced chemical equation. Outline the problem under the balanced equation. Calculate molar masses from atomic masses.

aluminum + titanium tetrachloride → aluminum chloride + titanium

4 Al(s) + 3 TiCl₄(l) → 4 AlCl₃(s) + 3 Ti(s)

$$2.00 \text{ mol TiCl}_4 \times \dfrac{3 \text{ mol Ti}}{3 \text{ mol TiCl}_4} \rightarrow \underline{} \text{ mol Ti}$$

MM = 47.9 g/mol

mass (g) Ti = $(\dfrac{47.9 \text{ g Ti}}{1 \text{ mol Ti}})(\dfrac{3 \text{ mol Ti}}{3 \text{ mol TiCl}_4})(2.00 \text{ mol TiCl}_4) = 95.8$ g Ti

(b) Find: percent yield (%) = ?

Given: actual yield = 71.0 g Ti

Known: theoretical yield = 95.8 g Ti

Solution: $\left(\dfrac{71.0 \text{ g actual}}{95.8 \text{ g theoretical}}\right)(100) = 74.1\%$

11.39 limiting reactant, O_2; 5.15 g CO_2 theoretical yield

Outline:

$CH_4(g) \; + \; 2\,O_2(g) \; \rightarrow \; CO_2(g) \; + \; 2\,H_2O(g)$

10.0 g   7.50 g   ? g

MM: 16.0 g/mol   32.0 g/mol   44.0 g/mol

0.625 mol   0.234 mol   (moles given)

Hypothesis: CH_4 is the limiting reactant.

Test: $(0.625 \text{ mol } CH_4)\left(\dfrac{2 \text{ mol } O_2}{1 \text{ mol } CH_4}\right) = 1.25$ mol O_2 needed if CH_4 is the limiting reactant. The number of moles of oxygen given is much less than the number of moles needed.

Conclusion: O_2 is the limiting reactant (LR).

Calculate the theoretical yield.

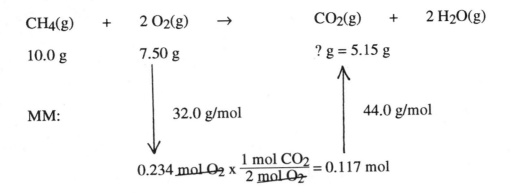

11.41 (a) LR is NH3   (b) 149 g (NH4)3PO4 theoretical yield

(a) Outline:

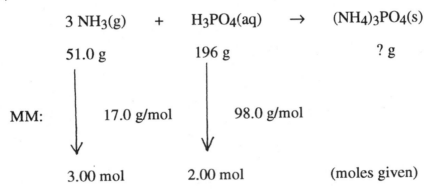

Hypothesis: NH3 is the limiting reactant.

Test: $(3.00 \text{ mol NH}_3)(\dfrac{1 \text{ mol H}_3\text{PO}_4}{3 \text{ mol NH}_3}) = 1.00$ mol H3PO4 needed if NH3 is the limiting reactant. The number of moles of H3PO4 given is more than the number of moles needed.

Conclusion: NH3 is the limiting reactant (LR).

(b) Calculate the theoretical yield.

Chapter 11: Calculations Involving Chemical Equations

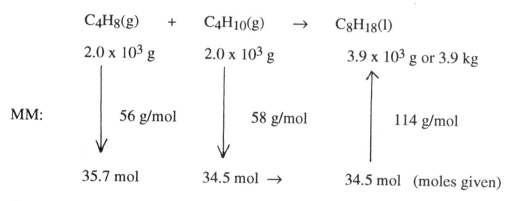

$$3\ NH_3(g)\ +\ H_3PO_4(aq)\ \rightarrow\ (NH_4)_3PO_4(s)$$

51.0 g   196 g   ? g = 149 g

MM = 17.0 g/mol   MM = 149 g/mol

$$3.00\ mol\ NH_3 \times \frac{1\ mol\ (NH_4)_3PO_4}{3\ mol\ NH_3} = 1.00\ mol$$

11.43 theoretical yield 3.9×10^3 g or 3.9 kg

Note that 2.0 kg is 2.0×10^3 g.

Outline:

$$C_4H_8(g)\ +\ C_4H_{10}(g)\ \rightarrow\ C_8H_{18}(l)$$

2.0×10^3 g 2.0×10^3 g 3.9×10^3 g or 3.9 kg

MM: 56 g/mol   58 g/mol   114 g/mol

35.7 mol   34.5 mol →   34.5 mol (moles given)

Conclusion: Since the mole ratio for the reactants is 1:1, C_4H_{10} is the limiting reactant.

11.45 (a) CCl_4 is LR.   (b) 39.3 g CCl_2F_2 (c) 11 g SbF_3 excess

(a)   Outline:

3 CCl$_4$(l) + 2 SbF$_3$(s) → 3 CCl$_2$F$_2$(g) + 2 SbCl$_3$(s)

50.0 g   50.0 g

MM: 154 g/mol   179 g/mol

0.325 mol   0.279 mol   (moles given)

Hypothesis: CCl$_4$ is the limiting reactant.

Test: (0.325 mol CCl$_4$)($\dfrac{2 \text{ mol SbF}_3}{3 \text{ mol CCl}_4}$) = 0.217 mol SbF$_3$ needed if CCl$_4$ is LR. The number of moles of SbF$_3$ given is more than the number of moles needed.

Conclusion: CCl$_4$ is LR.

(b) Calculate the theoretical yield of CCl$_2$F$_2$.

3 CCl$_4$(l) + 2 SbF$_3$(s) → 3 CCl$_2$F$_2$(g) + 2 SbCl$_3$(s)

50.0 g   ? g = 39.3 g

MM = 154 g/mol   MM = 121 g/mol

0.325 mol CCl$_4$ x $\dfrac{3 \text{ mol CCl}_2\text{F}_2}{3 \text{ mol CCl}_4}$ = 0.325 mol CCl$_2$F$_2$

Theoretical yield of CCl$_2$F$_2$ is 39.3 g.

(c) Find the mass of excess SbF$_3$.

Since 0.217 moles were needed and 0.279 moles were given, the difference (0.062 moles) is the amount in excess.

mass (g) SbF$_3$ = ($\dfrac{179 \text{ g SbF}_3}{1 \text{ mol SbF}_3}$)(0.062 mol SbF$_3$) = 11 g SbF$_3$

11.47 1.46 g C$_2$H$_5$Cl

Find: mass (g) C_2H_5Cl = ?

Given: 1.10 mL TEL per liter of gasoline

Known: density of TEL = 1.66 g/mL
$\dfrac{4 \text{ mol } C_2H_5Cl}{1 \text{ mol TEL}}$ from balanced equation

First determine mass of TEL needed for 1 liter of gasoline. Then use the equation to calculate the mass of ethyl chloride needed.

mass (g) TEL = $(\dfrac{1.66 \text{ g TEL}}{1 \text{ mL TEL}})(1.10 \text{ mL TEL})$ = 1.83 g TEL

Outline:

$4 \; C_2H_5Cl(l) \quad + \quad 4 \; NaPb(s) \quad \rightarrow \quad (C_2H_5)_4Pb(l) \quad + \quad 4 \; NaCl(s) \quad + \quad 3 \; Pb(s)$

? g   1.83 g

MM = 64.6 g/mol   MM = 323 g/mol

0.0227 mol $\leftarrow$ $\dfrac{4 \text{ mol } C_2H_5Cl}{1 \text{ mol TEL}}$ x 0.005 67 mol

mass (g) C_2H_5Cl = $(\dfrac{64.6 \text{ g } C_2H_5Cl}{1 \text{ mol } C_2H_5Cl})(\dfrac{4 \text{ mol } C_2H_5Cl}{1 \text{ mol TEL}})(\dfrac{1 \text{ mol TEL}}{323 \text{ g TEL}})(1.83 \text{ g TEL})$ = 1.46 g C_2H_5Cl

11.49 (a) 1.02×10^3 g C_2H_5OH theoretical yield   (b) 61.3% yield

(a) Outline the problem under the balanced equation. Calculate molar masses from atomic masses.

$C_6H_{12}O_6(s) \quad \rightarrow \quad 2 \; C_2H_5OH(l) \quad + \quad 2 \; CO_2(g)$

2000 g   ? g = 1.02×10^3 g

MM = 180. g/mol   MM = 46.0 g/mol

11.1 mol x $\dfrac{2 \text{ mol } C_2H_5OH}{1 \text{ mol } C_6H_{12}O_6}$ = 22.2 mol

(b) Find: percent yield (%) = ?

Given: 625 g actual yield

Known: 1.02×10^3 g theoretical yield

Solution: $\left(\dfrac{625 \text{ g actual}}{1.02 \times 10^3 \text{ g theoretical}}\right)(100) = 61.3\%$

11.51 (a) H_2 is the LR   (b) 51.5 g CH_3OH

(a) Outline:

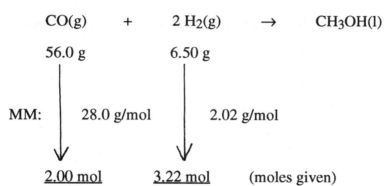

Hypothesis: CO is the limiting reactant.

Test: $\left(\dfrac{2 \text{ mol } H_2}{1 \text{ mol CO}}\right)(2.00 \text{ mol CO}) = 4.00$ mol H_2 needed if CO is the limiting reactant. The number of moles of H_2 given is less than the number of moles needed.

Conclusion: H_2 is LR.

(b) Outline:

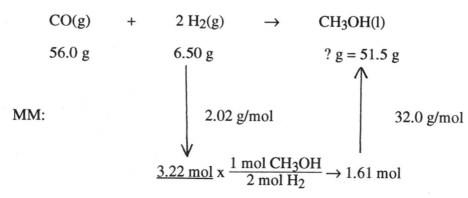

11.53 (a) 23.4 mol HNO$_3$ (b) 287 g H$_2$SO$_4$

(a) Outline:

$$H_2S(g) + 8\ HNO_3(aq) \rightarrow H_2SO_4(aq) + 8\ NO_2(g) + 4\ H_2O(l)$$

100. g

↓ MM = 34.1 g/mol

2.93 mol × $\dfrac{8\ \text{mol HNO}_3}{1\ \text{mol H}_2\text{S}}$ = 23.4 mol HNO$_3$

(b) Outline:

$$H_2S(g) + 8\ HNO_3(aq) \rightarrow H_2SO_4(aq) + 8\ NO_2(g) + 4\ H_2O(l)$$

100. g   ? g = 287 g

↓ MM = 34.1 g/mol   ↑ MM = 98.1 g/mol

2.93 mol × $\dfrac{1\ \text{mol H}_2\text{SO}_4}{1\ \text{mol H}_2\text{S}}$ = 2.93 mol

Chapter 12:
The Gaseous State

12.1 by volume: 78% nitrogen, 21% oxygen, 1% argon, <1% carbon dioxide and other trace gases

12.3 During thunderstorms, nitrogen combines with oxygen to form nitrogen oxides. These nitrogen oxides are washed out of the atmosphere by rainfall and fertilize plants.

12.5 A bullet travels more slowly through water than through air due to the difference in densities of these two fluids. There are very few molecules in a given volume of air, so there is little resistance to the movement of a bullet. In water, however, the molecules are much closer together and thus interfere with the movement of the bullet.

12.7 The sense of smell depends on the tendency of a gas to fill its container. Gaseous molecules move rapidly through space until they encounter an obstacle. When an animal is swimming underwater, its sense of smell is practically ineffective because the molecules having odor are stopped by the water before they can encounter the animal's nose.

12.9 A volume of gas cannot be measured by pouring it into a graduated cylinder because the gaseous molecules would completely fill the cylinder to fill the room and then escape from the cylinder to fill the room.

12.11 molecular velocity: $CO_2 < Ar < O_2 < N_2 < CH_4$.

Molecules of smaller mass have greater average velocity.

12.13 Gasoline molecules evaporated from the pan and moved around in the garage to fill the garage. When gas molecules in appropriate concentration reached the pilot light of the space heater, ignition occurred.

12.15 increasing average molecular velocity: $CO_2 < Ar < O_2 < N_2$

12.17 Pressure is force per unit area. Pressure in a gas is caused by collisions of gas particles with the walls of the container.

12.19 574 torr

Barometric pressure was greater than the pressure of the gas because it pushed down harder on the column of the mercury. Therefore the pressure in the cylinder is less than barometric by 132 torr: 706 torr - 132 torr = 574 torr.

12.21 152.8 atm, 1.161 x 10⁵ torr

Find: pressure (atm) = ?, pressure (torr) = ?

Given: 2245 psi

Known: $\dfrac{1 \text{ atm}}{14.69 \text{ psi}}$, $\dfrac{1 \text{ atm}}{760 \text{ torr}}$

Solution: P (atm) = $(\dfrac{1 \text{ atm}}{14.69 \text{ psi}})$(2245 psi) = 152.8 atm

P (torr) = $(\dfrac{760 \text{ torr}}{1 \text{ atm}})$(152.8 atm) = 1.161 x 10⁵ torr

12.23 0.149 atm

Find: P (atm) = ?

Given: 113 torr

Known: $\dfrac{1 \text{ atm}}{760 \text{ torr}}$

Solution: P (atm) = $(\dfrac{1 \text{ atm}}{760 \text{ torr}})$(113 torr) = 0.149 atm

12.25 0.933 atm

Find: P_2 (atm) = ?

Given: P_1 = 2.35 atm, V_1 = 1.25 L, V_2 = 3.15 L

Known: $P_2 = \dfrac{P_1 V_1}{V_2}$ or P↓, V↑

Solution: Using algebra, P_2 (atm) = $\dfrac{(2.35 \text{ atm})(1.25 \text{ L})}{(3.15 \text{ L})}$ = 0.933 atm

12.27 63.8 mL

Find: V_2 (mL) = ?

130 Student's Solutions Manual

Given: $V_1 = 45.6$ mL, $P_1 = 0.945$ atm, $P_2 = 0.675$ atm
Known: $V_2 = \frac{P_1 V_1}{P_2}$ or $P\downarrow, V\uparrow$

Solution: Using dimensional analysis, final volume = initial volume x volume ratio greater than 1
$V_2 (L) = (45.6 \text{ mL})(\frac{0.945 \text{ atm}}{0.675 \text{ atm}}) = 63.8$ mL

12.29 7.59 L

Find: $V_2 (L) = ?$
Given: $V_1 = 6.35$ L, $T_1 = 272$ K, $T_2 = 325$ K
Known: $\frac{V_1}{T_1} = \frac{V_2}{T_2}$ or $V\uparrow, T\uparrow$
Solution: Using algebra, $V_2 (L) = \frac{V_1 T_2}{T_1} = \frac{(6.35 \text{ L})(325 \text{ K})}{(272 \text{ K})} = 7.59$ L

12.31 36 mL

Find: $V_2 (L) = ?$
Given: $V_1 = 125$ mL L, $T_1 = 26 + 273 = 299$ K, $T_2 = -186 + 273 = 87$ K
Known: $\frac{V_1}{T_1} = \frac{V_2}{T_2}$ or $V\uparrow, T\uparrow$

Solution: Using dimensional analysis, volume will decrease as temperature decreases. Thus, multiply by fraction < 1. $V_2 (L) = (125 \text{ mL})(\frac{87 \text{ K}}{299 \text{ K}}) = 36$ mL

12.33 13.0 atm

Find: $P_2 (atm) = ?$
Given: $P_1 = 10.2$ atm, $T_1 = 21 + 273 = 294$ K, $T_2 = 103 + 273 = 376$ K
Known: $\frac{P_1}{T_1} = \frac{P_2}{T_2}$ or $P\uparrow, T\uparrow$

Solution: Using algebra, P_2 (atm) $= \dfrac{P_1 T_2}{T_1} = \dfrac{(10.2 \text{ atm})(376 \cancel{K})}{(294 \cancel{K})} = 13.0$ atm

12.35 24.3 atm

Find: P_2 (atm) = ?

Given: $P_1 = 22.3$ atm, $T_1 = 2.0 + 273.2 = 275.2$ K, $T_2 = 27 + 273 = 3.00 \times 10^2$ K

Known: $\dfrac{P_1}{T_1} = \dfrac{P_2}{T_2}$ or P↑, T↑

Solution: Using dimensional analysis, pressure will increase as temperature increases. Thus, multiply by fraction > 1. P_2 (atm) $= (22.3 \text{ atm})(\dfrac{300. \cancel{K}}{275.2 \cancel{K}}) = 24.3$ atm

12.37 STP refers to a standard temperature and pressure, which has been defined as 273 K and 1 atm.

12.39 1.37 L

Find: V_2 (L) at STP = ?

Given: $V_1 = 1.00$ L, $T_1 = 25 + 273 = 298$ K, $P_1 = 1.50$ atm
$T_2 = 273$ K, $P_2 = 1$ atm (exactly)

Known: $\dfrac{P_1 V_1}{T_1} = \dfrac{P_2 V_2}{T_2}$ or P↓, V↑ and T↑, V↑

Solution: Using algebra, V_2 (L) $= \dfrac{P_1 V_1 T_2}{P_2 T_1} = \dfrac{(1.50 \cancel{\text{atm}})(1.00 \text{ L})(273 \cancel{K})}{(1 \cancel{\text{atm}})(298 \cancel{K})} = 1.37$ L

12.41 5.01×10^3 torr

Find: P_2 (torr) = ?

Given: $V_1 = 0.538$ L, $T_1 = 96 + 273 = 369$ K, $P_1 = 695$ torr

Known: $T_2 = 345 + 273 = 618$ K, $V_2 = 0.125$ L

$\frac{P_1V_1}{T_1} = \frac{P_2V_2}{T_2}$ or $P\uparrow$, $V\downarrow$ and $T\uparrow$, $P\uparrow$

Solution: Using dimensional analysis, $P_2 = P_1 \times$ volume ratio $\times$ temperature ratio. Temperature ratio > 1, because temperature is increasing; volume ratio > 1, because volume is decreasing. P_2 (torr) = (695 torr)$(\frac{0.538 \text{ L}}{0.125 \text{ L}})(\frac{618 \text{ K}}{369 \text{ K}})$ = 5.01 x 10³ torr

12.43 7.73 x 10⁻³ L

Find: V_2 (L) = ?

Given: $V_1 = 10.0$ L, $T_1 = 135$ K, $P_1 = 3.50$ x 10⁻³ torr

$T_2 = 298$ K, $P_2 = 10.0$ torr

Known: $\frac{P_1V_1}{T_1} = \frac{P_2V_2}{T_2}$ or $P\downarrow$, $V\uparrow$ and $T\uparrow$, $V\uparrow$

Solution: Using algebra, V_2 (L) = $\frac{P_1V_1T_2}{P_2T_1} = \frac{(3.50 \times 10^{-3} \text{ torr})(10.0 \text{ L})(298 \text{ K})}{(10.0 \text{ torr})(135 \text{ K})}$ = 7.73 x 10⁻³ L

12.45 1.43 mol

Find: # mol gas = ?

Given: 0.325 mol at 6.50 L; 28.5 L

Known: $n\uparrow$, $V\uparrow$

Solution: n (mol) = (0.325 mol)$(\frac{28.5 \text{ L}}{6.50 \text{ L}})$ = 1.43 mol

12.47 1.98 L

Find: V (L) = ?

Given: 28.0 g N_2 occupies 1.00 L; 4.00 g H_2

Known: n↑, V↑; MM-H$_2$ = 2.02 g/mol, MM-N$_2$ = 28.0 g/mol

Solution: Calculate moles of N$_2$ and of H$_2$

$$\text{mol N}_2 = \left(\frac{1 \text{ mol N}_2}{28.0 \text{ g N}_2}\right)(28.0 \text{ g N}_2) = 1.00 \text{ mol N}_2$$

$$\text{mol H}_2 = \left(\frac{1 \text{ mol H}_2}{2.02 \text{ g H}_2}\right)(4.00 \text{ g H}_2) = 1.98 \text{ mol H}_2$$

volume H$_2$ = volume N$_2$ × mole ratio = $(1.00 \text{ L})\left(\frac{1.98 \text{ mol}}{1.00 \text{ mol}}\right)$ = 1.98 L

12.49 5.15 L

Find: V (L) = ?

Given: 0.230 mol CH$_4$ at STP

Known: 22.4 L/mol at STP

Solution: V (L) = $\left(\frac{22.4 \text{ L}}{1 \text{ mol}}\right)(0.230 \text{ mol})$ = 5.15 L

12.51 12.6 L

Find: V (L) = ?

Given: 18.0 g O$_2$ at STP

Known: 22.4 L/mol at STP; MM-O$_2$ = 32.0 g/mol

Solution: V (L) = $\left(\frac{22.4 \text{ L}}{1 \text{ mol}}\right)\left(\frac{1 \text{ mol}}{32.0 \text{ g}}\right)(18.0 \text{ g})$ = 12.6 L

12.53 1.67 × 10^5 mol

Find: mol He = ?

Given: 3.75 × 10^3 m^3; $\frac{1 \text{ m}^3}{1000 \text{ L}}$

Known: 22.4 L/mol at STP

Solution: mol = $(\frac{1 \text{ mol}}{22.4 \text{ L}})(\frac{1000 \text{ L}}{1 \text{ m}^3})(3.75 \times 10^3 \text{ m}^3) = 1.67 \times 10^5$ mol

12.55 4.4 L

Find: V (L C_3H_8) = ?

Given: 22 L oxygen; same temperature and pressure

Known: Gas volumes are proportional to moles at same temperature and pressure

$C_3H_8(g) + 5 O_2(g) \rightarrow 3 CO_2(g) + 4 H_2O(l)$ ∴ $\frac{5 \text{ L } O_2}{1 \text{ L } C_3H_8}$

Solution: V (L C_3H_8) = $(\frac{1 \text{ L } C_3H_8}{5 \text{ L } O_2})(22 \text{ L } O_2) = 4.4$ L C_3H_8

12.57 13 L

Find: V (L SO_2) = ?

Given: 6.5 L CO_2

Known: Gas volumes are proportional to moles at same temperature and pressure

$CS_2(l) + 3 O_2(g) \rightarrow CO_2(g) + 2 SO_2(g)$ ∴ $\frac{2 \text{ L } SO_2}{1 \text{ L } CO_2}$

Solution: V (L SO_2) = $(\frac{2 \text{ L } SO_2}{1 \text{ L } CO_2})(6.5 \text{ L } CO_2) = 13$ L SO_2

12.59 3.11 atm

Find: P (atm) = ?

Given: n = 0.128 mol, V = 1.00 L, T = 23 + 273 = 296 K

Known: PV = nRT, R = 0.0821 $\frac{\text{L·atm}}{\text{mol·K}}$

Solution: $P \text{ (atm)} = \dfrac{nRT}{V} = \dfrac{(0.128 \text{ mol})(0.0821 \frac{L \cdot atm}{mol \cdot K})(296 \text{ K})}{(1.00 \text{ L})} = 3.11 \text{ atm}$

12.61 0.370 mol

Find: n (mol) = ?

Given: V = 3.42 L, P = 2.65 atm, T = 298 K

Known: PV = nRT, R = 0.0821 $\frac{L \cdot atm}{mol \cdot K}$

Solution: $n = \dfrac{PV}{RT} = \dfrac{(2.65 \text{ atm})(3.42 \text{ L})}{(0.0821 \frac{L \cdot atm}{mol \cdot K})(298 \text{ K})} = 0.370 \text{ mol}$

12.63 1.4 g

Find: mass (g) = ?

Given: V = 2.2 L, T = 28 + 273 = 301 K, P = 755 torr

Known: PV = nRT, R = 0.0821 $\frac{L \cdot atm}{mol \cdot K}$

$P \text{ (atm)} = (\dfrac{1 \text{ atm}}{760 \text{ torr}})(755 \text{ torr}) = 0.993 \text{ atm}$

MM-CH_4 = 16.0 g/mol

Solution: Solve for n using algebra, then convert moles to grams using molar mass.

$n = \dfrac{PV}{RT} = \dfrac{(0.993 \text{ atm})(2.2 \text{ L})}{(0.0821 \frac{L \cdot atm}{mol \cdot K})(301 \text{ K})} = 0.088 \text{ mol}$

mass (g) = $(\dfrac{16.0 \text{ g}}{1 \text{ mol}})(0.088 \text{ mol}) = 1.4 \text{ g}$

12.65 4.69 L

Find: V (L) = ?

Given: 3.20 g NH$_3$, P = 733 torr, T = 293 K

Known: PV = nRT, R = 0.0821 $\frac{\text{L·atm}}{\text{mol·K}}$

P (atm) = ($\frac{1 \text{ atm}}{760 \text{ torr}}$)(733 torr) = 0.964 atm

MM-NH$_3$ = 17.0 g/mol, n (mol) = ($\frac{1 \text{ mol}}{17.0 \text{ g}}$)(3.20 g) = 0.188 mol

Solution: V (L) = $\frac{nRT}{P}$ = $\frac{(0.188 \text{ mol})(0.0821 \frac{\text{L·atm}}{\text{mol·K}})(293 \text{ K})}{(0.964 \text{ atm})}$ = 4.69 L

12.67 44.1 g/mol

Find: MM (g/mol) = ?

Given: 0.234 g, 119 mL at STP

Known: V = 119 mL = 0.119 L, T = 273 K, P = 1 atm (exactly)

PV = nRT, R = 0.0821 $\frac{\text{L·atm}}{\text{mol·K}}$

Solution: Solve for moles using ideal gas equation, then solve for molar mass.

n (mol) = $\frac{PV}{RT}$ = $\frac{(1 \text{ atm})(0.119 \text{ L})}{(0.0821 \frac{\text{L·atm}}{\text{mol·K}})(273 \text{ K})}$ = 0.005 31 mol

MM (g/mol) = $\frac{0.234 \text{ g}}{0.005 \text{ 31 mol}}$ = 44.1 g/mol

An alternative solution is to calculate moles from the known ratio 22.4 L/mol of gas at STP:

n (mol) = ($\frac{1 \text{ mol}}{22.4 \text{ L}}$)(0.119 L) = 0.005 31 mol

Chapter 12: The Gaseous State   137

12.69   6.24 L

Find: V (L) = ?

Given: MM-CH_4 = 16.0 g/mol, 3.88 g, P = 719 torr, T = 23 + 273 = 296 K

Known: PV = nRT, R = 0.0821 $\frac{L \cdot atm}{mol \cdot K}$

P (atm) = ($\frac{1 \text{ atm}}{760 \text{ torr}}$)(719 torr) = 0.946 atm

n (mol) = ($\frac{1 \text{ mol}}{16.0 \text{ g}}$)(3.88 g) = 0.243 mol

Solution: V (L) = $\frac{nRT}{P}$ = $\frac{(0.243 \text{ mol})(0.0821 \frac{L \cdot atm}{mol \cdot K})(296 \text{ K})}{(0.946 \text{ atm})}$ = 6.24 L

12.71   D: CH_4 < O_2 < HCl < SO_2 < HI

Density is directly proportional to molar mass. Therefore, as molar mass increases at a given temperature and pressure, the density will increase. MM (g/mol) : O_2 = 32.0, CH_4 = 16.0, HCl = 36.5, HI = 128, SO_2 = 64.0.

12.73   17.0 g/mol

Find: MM (g/mol) = ?

Given: D = 0.759 g/L at STP

Known: D = MM $\frac{P}{RT}$, R = 0.0821 $\frac{L \cdot atm}{mol \cdot K}$, P = 1 atm (exactly), T = 273 K

Solution: MM (g/mol) = $\frac{DRT}{P}$ = $\frac{(0.759 \text{ g/L})(0.0821 \frac{L \cdot atm}{mol \cdot K})(273 \text{ K})}{(1 \text{ atm})}$ = 17.0 g/mol

An alternative solution utilizes the known ratio 22.4 L/mol of gas at STP:

MM (g/mol) = ($\frac{0.759 \text{ g}}{L}$)($\frac{22.4 \text{ L}}{mol}$) = 17.0 g/mol

12.75 1.16 g/mL

Find: D_2 (g/L) = ?

Given: P_2 = 1.00 atm, T_2 = 295 K, D_1 = 1.25 g/L at STP

Known: T_1 = 273 K, P_1 = 1 atm. As T↑, D↓ if pressure is the same.

Solution: Using dimensional analysis, multiply initial density by a temperature ratio.
D_2 (g/L) = (1.25 g/L)($\frac{273 \text{ K}}{295 \text{ K}}$) = 1.16 g/mL

12.77 (a) 619 torr (b) 0.001 38 mol

(a) Find: P_{O_2} (torr) = ?

Given: P_{bar} = 643 torr, P_{H_2O} = 23.8 torr

Known: $P_T = P_{bar} = P_{O_2} + P_{H_2O}$

Solution: P_{O_2} (torr) = $P_{bar} - P_{H_2O}$ = 643 torr - 23.8 torr = 619 torr

(b) Find: n (mol) = ?

Given: P_{O_2} = 619 torr, V = 41.6 mL, T = 25 + 273 = 298 K

Known: $\frac{1 \text{ L}}{1000 \text{ mL}}$, $\frac{1 \text{ atm}}{760 \text{ torr}}$, PV = nRT, R = 0.0821 $\frac{\text{L·atm}}{\text{mol·K}}$

Solution: V (L) = ($\frac{1 \text{ L}}{1000 \text{ mL}}$)(41.6 mL) = 0.0416 L

P (atm) = ($\frac{1 \text{ atm}}{760 \text{ torr}}$)(619 torr) = 0.814 atm

n = $\frac{PV}{RT}$ = $\frac{(0.814 \text{ atm})(0.0416 \text{ L})}{(0.0821 \frac{\text{L·atm}}{\text{mol·K}})(298 \text{ K})}$ = 0.001 38 mol

Chapter 12: The Gaseous State   139

12.79 9.9 L

Find: V (L) = ?

Given: 0.15 mol He and 0.25 mol Ne, T = 28 + 273 = 301 K, P_T = 1.00 atm

Known: PV = nRT, R = 0.0821 $\frac{L \cdot atm}{mol \cdot K}$

Solution: Combine moles to obtain total moles, and then solve the ideal gas equation for V

total moles = 0.15 mol He + 0.25 mol Ne = 0.40 mol gas

$$V (L) = \frac{nRT}{P} = \frac{(0.40 \text{ mol})(0.0821 \frac{L \cdot atm}{mol \cdot K})(301 \text{ K})}{(1.00 \text{ atm})} = 9.9 \text{ L}$$

12.81 0.0181 mol H_2O_2

Outline: 2 H_2O_2(aq) → 2 H_2O(l) + O_2(g)

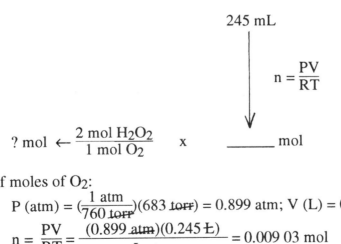

number of moles of O_2:

P (atm) = $(\frac{1 \text{ atm}}{760 \text{ torr}})$(683 torr) = 0.899 atm; V (L) = 0.245 L; T = 297 K

$$n = \frac{PV}{RT} = \frac{(0.899 \text{ atm})(0.245 \text{ L})}{(0.0821 \frac{L \cdot atm}{mol \cdot K})(297 \text{ K})} = 0.009\ 03 \text{ mol}$$

number of moles of H_2O_2 = $(\frac{2 \text{ mol } H_2O_2}{1 \text{ mol } O_2})$(0.009 03 mol O_2) = 0.0181 mol

12.83 194 g HgO

Write a balanced equation and outline the problem.

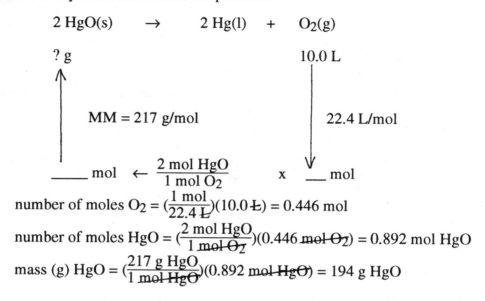

number of moles $O_2 = (\frac{1 \text{ mol}}{22.4 \text{ L}})(10.0 \text{ L}) = 0.446$ mol

number of moles HgO $= (\frac{2 \text{ mol HgO}}{1 \text{ mol } O_2})(0.446 \text{ mol } O_2) = 0.892$ mol HgO

mass (g) HgO $= (\frac{217 \text{ g HgO}}{1 \text{ mol HgO}})(0.892 \text{ mol HgO}) = 194$ g HgO

12.85 12 g NH$_4$Cl

This is a limiting reactant problem. However, equal volumes of gases at the same temperature and pressure contain the same number of moles. Since the mole ratio for the reactants in the equation is 1 mol NH$_3$/1 mol HCl, exactly the correct amounts of each gas have been mixed. So both gases are equally limiting. You may use either gas to solve the problem.

Outline: NH$_3$(g) + HCl(g) → NH$_4$Cl(s)

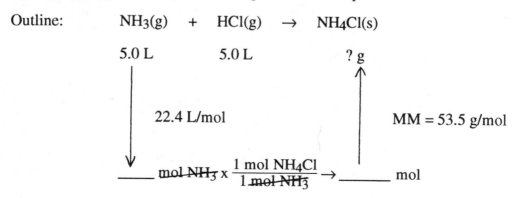

number of moles NH$_3$ = ($\frac{1 \text{ mol}}{22.4 \text{ L}}$)(5.0 L) = 0.22 mol

number of moles NH$_4$Cl = ($\frac{1 \text{ mol NH}_4\text{Cl}}{1 \text{ mol NH}_3}$)(0.22 mol NH$_3$) = 0.22 mol NH$_4$Cl

mass (g) NH$_4$Cl = ($\frac{53.5 \text{ g NH}_4\text{Cl}}{1 \text{ mol NH}_4\text{Cl}}$)(0.22 mol NH$_4$Cl) = 12 g NH$_4$Cl

12.87 1.42 x 10^4 g Cu$_2$S

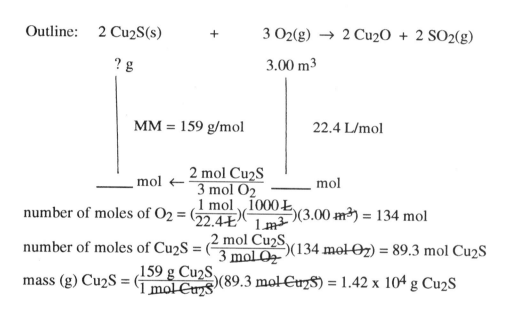

Outline: 2 Cu$_2$S(s) + 3 O$_2$(g) → 2 Cu$_2$O + 2 SO$_2$(g)

number of moles of O$_2$ = ($\frac{1 \text{ mol}}{22.4 \text{ L}}$)($\frac{1000 \text{ L}}{1 \text{ m}^3}$)(3.00 m^3) = 134 mol

number of moles of Cu$_2$S = ($\frac{2 \text{ mol Cu}_2\text{S}}{3 \text{ mol O}_2}$)(134 mol O$_2$) = 89.3 mol Cu$_2$S

mass (g) Cu$_2$S = ($\frac{159 \text{ g Cu}_2\text{S}}{1 \text{ mol Cu}_2\text{S}}$)(89.3 mol Cu$_2$S) = 1.42 x 10^4 g Cu$_2$S

12.89 2.66 g BaCO$_3$

Outline: Ba(OH)$_2$(aq) + CO$_2$(g) → BaCO$_3$(s) + H$_2$O(l)

number of moles of CO_2:   V (L) = 0.325 L; T = 21 + 273 = 294 K

$$n = \frac{PV}{RT} = \frac{(1.00 \text{ atm})(0.325 \text{ L})}{(0.0821 \frac{\text{L·atm}}{\text{mol·K}})(294 \text{ K})} = 0.0135 \text{ mol}$$

number of moles of $BaCO_3$ = $(\frac{1 \text{ mol } BaCO_3}{1 \text{ mol } CO_2})(0.0135 \text{ mol } CO_2)$ = 0.0135 mol $BaCO_3$

mass (g) $BaCO_3$ = $(\frac{197 \text{ g } BaCO_3}{1 \text{ mol } BaCO_3})(0.0135 \text{ mol } BaCO_3)$ = 2.66 g $BaCO_3$

12.91 119 L

Outline: 2 NaCl(l) + electric current → 2 Na(l) + Cl_2(g)

425 g   ? L

MM = 58.5 g/mol $V = \frac{nRT}{P}$

_____ mol NaCl × $\frac{1 \text{ mol } Cl_2}{2 \text{ mol NaCl}}$ → _____ mol

number of moles of NaCl = $(\frac{1 \text{ mol}}{58.5 \text{ g}})(425 \text{ g})$ = 7.26 mol

number of moles of Cl_2 = $(\frac{1 \text{ mol } Cl_2}{2 \text{ mol NaCl}})(7.26 \text{ mol NaCl})$ = 3.63 mol Cl_2

$$V (L) = \frac{nRT}{P} = \frac{(3.63 \text{ mol})(0.0821 \frac{\text{L·atm}}{\text{mol·K}})(125 + 273 \text{ K})}{(1.00 \text{ atm})} = 119 \text{ L}$$

12.93 29.3 L/mol

Find: molar volume (L/mol) = ?

Given: T = 24 + 273 = 297 K, P = 632 torr

Known: $\frac{1 \text{ atm}}{760 \text{ torr}}$, PV = nRT, R = 0.0821 $\frac{\text{L·atm}}{\text{mol·K}}$

Solution: P (atm) = $(\frac{1 \text{ atm}}{760 \text{ torr}})(632 \text{ torr})$ = 0.832 atm

$$\frac{V}{n}\left(\frac{L}{mol}\right) = \frac{RT}{P} = \frac{(0.0821 \frac{L \cdot atm}{mol \cdot K})(297 \text{ K})}{(0.832 \text{ atm})} = 29.3 \frac{L}{mol}$$

12.95 0.0931 g

Find: mass (g) = ?
Given: V = 1.35 L, T = 24 + 273 = 297 K, P = 632 torr
Known: MM-H_2 = 2.02 g/mol, PV = nRT, R = 0.0821 $\frac{L \cdot atm}{mol \cdot K}$
Solution: P = $(\frac{1 \text{ atm}}{760 \text{ torr}})$(632 torr) = 0.832 atm

$$n \text{ (mol)} = \frac{PV}{RT} = \frac{(0.832 \text{ atm})(1.35 \text{ L})}{(0.0821 \frac{L \cdot atm}{mol \cdot K})(297 \text{ K})} = 0.0461 \text{ mol}$$

mass (g) = $(\frac{2.02 \text{ g}}{1 \text{ mol}})$(0.0461 mol) = 0.0931 g H_2

12.97 2.4 L

Since the volumes of the gases will be proportional to the moles of the gases, the coefficients in the equation can be read as the number of liters of each gas that will be produced: 6 L N_2 + 1 L O_2 + 12 L CO_2 + 10 L H_2O = 29 L gas. Thus, when 29 L of gas are produced, 12 L CO_2 are formed. A ratio must be used to calculate the volume of total gases that would contain 1.0 L CO_2.

V(L) gases = $(\frac{29 \text{ L gases}}{12 \text{ L } CO_2})$(1.0 L CO_2) = 2.4 L gases

12.99 0.156 g $KClO_3$

Find: mass (g) $KClO_3$ = ?

Given: balanced chemical equation, V = 46.8 mL = 0.0468 L, T = 15 + 273 = 288 K, P_{bar} = 748, P_{H_2O} = 12.8 torr

Known: $P_{O_2} = P_{bar} - P_{H_2O} = 748 - 12.8 = (735 \text{ torr})(\frac{1 \text{ atm}}{760 \text{ torr}}) = 0.967$ atm

$PV = nRT$, $R = 0.0821 \frac{L \cdot atm}{mol \cdot K}$

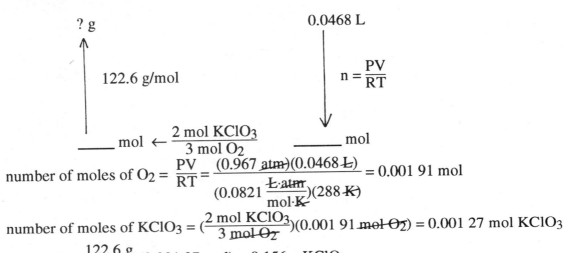

number of moles of $O_2 = \frac{PV}{RT} = \frac{(0.967 \text{ atm})(0.0468 \text{ L})}{(0.0821 \frac{L \cdot atm}{mol \cdot K})(288 \text{ K})} = 0.00191$ mol

number of moles of $KClO_3 = (\frac{2 \text{ mol } KClO_3}{3 \text{ mol } O_2})(0.00191 \text{ mol } O_2) = 0.00127$ mol $KClO_3$

mass (g) = $(\frac{122.6 \text{ g}}{1 \text{ mol}})(0.00127 \text{ mol}) = 0.156$ g $KClO_3$

12.101 2.46×10^{22} molecules

Find: number of molecules $H_2 = ?$

Given: V = 1.00 L, T = 25 + 273 = 298 K, P = 1.00 atm

Known: MM-H_2 = 2.02 g/mol, $\frac{6.02 \times 10^{23} \text{ molecules}}{1 \text{ mol}}$

$PV = nRT$, $R = 0.0821 \frac{L \cdot atm}{mol \cdot K}$

Solution: Solve for moles using the ideal gas equation, and then solve for number of molecules.

$n = \frac{PV}{RT} = \frac{(1.00 \text{ atm})(1.00 \text{ L})}{(0.0821 \frac{L \cdot atm}{mol \cdot K})(298 \text{ K})} = 0.0409$ mol

number of molecules = $(\frac{6.02 \times 10^{23} \text{ molecules}}{1 \text{ mol}})(0.0409 \text{ mol}) = 2.46 \times 10^{22}$ molecules

12.103   16.6 L

Find: V (L) = ?

Given: 36.3 g CO_2, T = 33 + 273 = 306 K, P = 1.25 atm

Known: MM-CO_2 = 44.0 g/mol, PV = nRT, R = 0.0821 $\frac{L \cdot atm}{mol \cdot K}$

Solution: Solve for moles of CO_2, and then use the ideal gas equation to solve for V(L).

number of moles of CO2 = $(\frac{1 \text{ mol}}{44.0 \text{ g}})(36.3 \text{ g})$ = 0.825 mol

$$V = \frac{nRT}{P} = \frac{(0.825 \text{ mol})(0.0821 \frac{L \cdot atm}{mol \cdot K})(306 \text{ K})}{(1.25 \text{ atm})} = 16.6 \text{ L}$$

Chapter 13:
Liquids, Solids, and Changes of State

13.1 The general properties of a liquid are lack of a definite shape, incompressible, flows readily, slow diffusion, and densities close to those of solids.

13.3 A liquid flows readily because its molecules move while remaining in contact with one another.

13.5 A liquid diffuses more slowly than a gas because its molecules are so close together that they inhibit the rapid movement possible for molecules in a gas. Many more collisions occur per unit time in a liquid, and many of these collisions inhibit forward motion.

13.7 dipole-dipole attractions, hydrogen bonds, and dispersion forces

13.9 Since HBr is a polar molecule, it has both dipole-dipole attractions and dispersion forces.

13.11 Dipole-dipole attractions are practically ineffective with gases because they are weak forces that are not significant at the great distances between gas molecules.

13.13 Polar liquids are generally volatile because dipole-dipole attractions are a weak force of attraction.

13.15 The molecular features necessary for hydrogen bonding are an H—F, H—O, or H—N bond.

13.17

[Diagram showing hydrogen bonding between two water molecules and between two ammonia molecules, labeled "hydrogen bonding"]

13.19 hydrophilic: a, ethyl alcohol, b, methyl amine, and c, ammonia. These molecules are hydrophilic because each has either an O—H or N—H bond and therefore can form hydrogen bonds with water molecules.

13.21 Nonpolar molecules such as C_5H_{12} are attracted to one another through temporary induced dipoles that result from the movement of electrons within the molecule. These intermolecular forces are called dispersion or London forces.

13.23 Since dispersion forces increase with size and molar mass, large molecules have strong dispersion forces. An example is octane, C_8H_{18}.

13.25 The rate of vaporization of a liquid increases with increasing temperature because a larger fraction of molecules have enough kinetic energy to escape to the vapor phase.

13.27 Dynamic equilibrium of a liquid and its vapor means that molecules are continually evaporating and condensing even though the total number of molecules in the vapor phase remains constant. Equilibrium vapor pressure of a liquid refers to the vapor pressure exerted by the liquid at equilibrium with its vapor.

13.29 Sublimation is the process of going directly from the solid state to the vapor state. Some substances that sublime are iodine, *p*-dichlorobenzene (moth crystals), and dry ice (solid carbon dioxide).

13.31 (a) benzene (b) ethyl ether (c) chloroform (d) toluene

(a) Both are nonpolar molecules, having only dispersion forces that increase with molar mass. Toluene has a larger molar mass than benzene. Since stronger intermolecular forces result in lower vapor pressures, toluene has the lower vapor pressure and benzene has the higher vapor pressure. (b) Both molecules have the same molar mass and therefore similar dispersion forces. Butyl alcohol can participate in hydrogen bonding, which results in stronger intermolecular forces and lower vapor pressure than ethyl ether. (c) Chloroform is polar but has a smaller molar mass (120 g/mol) than carbon tetrachloride (154 g/mol). Thus, carbon tetrachloride has stronger dispersion forces and a lower vapor pressure. (Even though dispersion forces are not very strong individually, they increase rapidly with molar mass. The increase in strength with molar mass is sufficient to make dispersion forces more important in this case than the dipole-dipole attractions in chloroform.) (d) Chlorobenzene has a greater molar mass than toluene. Chlorobenzene is also slightly polar. This combination results in stronger intermolecular forces and lower vapor pressure.

13.33 (a) toluene (b) butyl alcohol (c) carbon tetrachloride (d) chlorobenzene

The higher the vapor pressure, the lower the boiling point. Thus, the other member of each pair in problem 31 will have the higher boiling point.

13.35 2.3×10^3 kJ

Find: energy (kJ) = ?

Given: 1.00 L water, 0.998 g/mL

Known: 41 kJ/mol, MM-H$_2$O = 18.0 g/mol, 1.00 L = 1.00 × 10^3 mL

Solution: energy (kJ) = $(\frac{41 \text{ kJ}}{1 \text{ mol}})(\frac{1 \text{ mol}}{18.0 \text{ g}})(\frac{0.998 \text{ g}}{1 \text{ mL}})(1.00 \times 10^3 \text{ mL}) = 2.3 \times 10^3$ kJ

13.37 The electrostatic attractions between oppositely charged ions are much greater than the attractions between molecules. Therefore, ionic compounds have low vapor pressures and high boiling points.

13.39 NaCl has a higher normal boiling point than CH$_3$Cl because ionic compounds have stronger attractions between particles (ions) than do molecular compounds.

13.41 All three molecules are polar and have dipole-dipole attractions. In addition, there are dispersion forces. Because of its molar mass (128 g/mol), HI has stronger dispersion forces than either HBr (MM = 81 g/mol) or HCl (MM = 36.5 g/mol). As intermolecular forces increase in strength, it becomes more difficult to separate the molecules to form a vapor. Thus, the temperature required to produce a vapor pressure of one atmosphere is greater as the dispersion forces increase.

13.43 A crystalline solid is one that has a regular geometric pattern. Some examples are sodium chloride, diamond, quartz, and other gemstones.

13.45 A crystalline solid is distinguished from an amorphous solid by the physical properties of hardness and brittleness. Amorphous solids are flexible and elastic.

13.47 types of crystalline solids: ionic, molecular, network, and metallic

13.49 Sucrose forms a molecular crystal. The intermolecular forces in a sugar crystal are dipole-dipole attractions, hydrogen bonding, and dispersion forces.

13.51 Network solids have very high melting points compared with molecular crystals because the interparticle attractions are covalent bonds, which are much stronger than dipole-dipole attractions, hydrogen bonding, or dispersion forces and which, therefore, require more energy to disrupt for melting.

13.53 An aqueous sugar solution is a nonconductor of electricity because sugar is molecular. Without ions, there can be no conductivity in aqueous solutions.

13.55 (a) molecular   (b) ionic   (c) network   (d) molecular   (e) molecular   (f) ionic

Ice (a), iodine (d), and dry ice (e) are recognized as being molecular because the formula shows they are composed of nonmetals only. Fluorite (b) and marble (f) are ionic because their formulas show they are composed of ions. Quartz (c) is a network crystal—it is used to make glass. The formula itself is not obvious. You must recognize that quartz is another name for silica, and that most sands are silica. You know that sand is not easily melted. Since the formula is not ionic, it must therefore be network.

13.57 3.35×10^5 J or 335 kJ

Find: q (J) = ?

152 Student's Solutions Manual

Given: 1.00 kg ice at 0°C

Known: ΔH_{fus} = 335 J/g, 1 kg = 1000 g

Solution: q (J) = $(\frac{335 \text{ J}}{1 \text{ g}})(\frac{1000 \text{ g}}{1 \text{ kg}})(1.00 \text{ kg})$ = 3.35 x 10^5 J

This quantity of energy is more conveniently expressed in kJ.

q (kJ) = $(\frac{1 \text{ kJ}}{1000 \text{ J}})(3.35 \times 10^5 \text{ J})$ = 335 kJ

Note: This problem can also be solved by converting 1.00 kg H$_2$O into moles and using the molar heat of fusion, 6.03 kJ/mol, as is shown in some of the practice problem solutions.

13.59 636 kJ

Find: q (kJ) = ?

Given: 245 g water at 20°C, to steam at 100°C

Known: SH = 4.18 J/g·°C, ΔH_{vap} = 2.26 kJ/g, T = 80°C

Solution: There are two steps that must be calculated separately—heating the water from 20°C to 100°C, and then vaporizing the water at 100°C. The total energy is obtained by adding the two heats.

heating the water, q (kJ) = $(\frac{1 \text{ kJ}}{1000 \text{ J}})(\frac{4.18 \text{ J}}{1 \text{ g·°C}})(245 \text{ g})(80 \text{°C})$ = 81.9 kJ

boiling the water, q (kJ) = $(\frac{2.26 \text{ kJ}}{1 \text{ g}})(245 \text{ g})$ = 554 kJ

total energy, q (kJ) = 81.9 + 554 = 636 kJ

13.61 4.00 x 10^2 g

Find: mass (g) = ?

Given: condensing 50.0 g steam, cooling 50.0 g water from 100°C to 0°C

Known: ΔH_{cond} = 2.26 kJ/g, ΔH_{fus} = 335 J/g, SH-water = 4.18 J/g·°C, T = 100°C, 1 kJ = 1000 J

Solution: 4 steps: (1) heat lost on condensing, (2) heat lost on cooling, (3) total heat lost = total heat gained by ice, and (4) calculate mass of ice that will melt

condensing: q (kJ) = $(\frac{2.26 \text{ kJ}}{1 \text{ g}})(50.0 \text{ g})$ = 113 kJ

cooling: q (kJ) = $(\frac{1 \text{ kJ}}{1000 \text{ J}})(\frac{4.18 \text{ J}}{1 \text{ g} \cdot °C})(50.0 \text{ g})(100°C)$ = 20.9 kJ

total heat lost = total heat gained = 113 + 20.9 = 134 kJ

mass (g) = $(\frac{1 \text{ g}}{335 \text{ J}})(\frac{1000 \text{ J}}{1 \text{ kJ}})(134 \text{ kJ})$ = 4.00 x 10² g

13.63

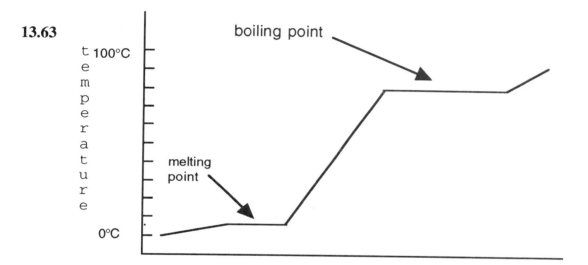

13.65 Surface tension is a result of intermolecular attractive forces. Increased temperature results in increased energy and velocity for the molecules of a liquid. As the energy of the molecules increases, intermolecular forces are relatively weaker—the molecules have too much energy to stay close to one another. Surface tension is the result of attractive forces pulling molecules together, but increased molecular velocity tends to pull molecules apart. The result is that surface tension decreases as temperature increases.

13.67 There are two reasons for the high viscosity of vegetable oils. There are very strong dispersion forces because the molecules are large. Additionally, the long carbon chains get tangled together and cannot flow freely over one another.

13.69 The properties of water that are most significant for its use as a coolant in industrial processes are its high specific heat and its large heat of vaporization. It takes a lot of heat to raise the temperature of a sample of water by even 1°C as compared to other substances. Furthermore, due to hydrogen bonding, it takes a very large amount of heat to boil water. Thus, water gets hot without boiling when used as an industrial coolant.

13.71 If the density of ice were greater than the density of liquid water, ice would form at the bottom of lakes and rivers and freeze up to the top. This would kill marine life. With ice being less dense than liquid water, the tops of lakes and rivers freezes. This top layer of ice insulates the remainder of the water. Thus, most lakes and rivers do not freeze solid and marine life can live.

13.73 Hydration means the attraction or bonding of water molecules to an ion or molecule.

13.75 A hydrate is a substance that contains water molecules in its crystalline structure. Hydrates have definite proportions of water as shown by their formulas and are therefore compounds. Examples include $CuSO_4 \cdot 5H_2O$, $MgSO_4 \cdot 7H_2O$, and $CaCl_2 \cdot 2H_2O$.

13.77 (a) $Na_2CO_3 \cdot 10H_2O$ (b) $MgSO_4 \cdot 7H_2O$ (c) $Ba(OH)_2 \cdot 8H_2O$

13.79 A deliquescent compound is a compound that absorbs enough moisture from the air to dissolve in it. Calcium chloride, $CaCl_2$, is an example of a deliquescent compound.

13.81 The boiling point of hydrogen peroxide, H_2O_2, is higher than the boiling point of water because there are two oxygen atoms in hydrogen peroxide that can participate in hydrogen bonding. The more hydrogen bonds that can form, the stronger the intermolecular forces holding the molecules together in the liquid state. Additionally, hydrogen peroxide molecules have a greater molar mass than water molecules with accordingly larger dispersion forces. Therefore, more energy, as represented by a higher boiling point, is required to reach the boiling point.

13.83 Water in a narrow glass tube has a meniscus because of attractive forces acting between the liquid and the surface of the glass. Since pentane (C_5H_{12}) is a nonpolar molecule, it is not attracted to the same kinds of substances as water, and no meniscus is formed in a narrow glass tube.

13.85 Ethyl alcohol has a much lower equilibrium vapor pressure at 25°C than does ethyl ether because ethyl alcohol has an OH group that can participate in hydrogen bonding. Hydrogen bonding is not possible in ether. In the absence of this strong attractive force, it is easier for ether molecules to vaporize than it is for alcohol molecules to vaporize. Thus, ethyl ether has a higher vapor pressure than ethyl alcohol.

13.87 100°C

    Find: T_{final} (°C) = ?

    Given: 175 kJ = 175 x 10^3 J, 0.100 kg water, T = 90°C

    Known: SH-water = 4.18 J/g·°C, 0.100 kg = 1.00 x 10^2 g, ΔH_{vap} = 2.26 kJ/g

Solution: Calculate the temperature change that will occur when 175 kJ of heat are added to the water. Then add the temperature change to the initial temperature to find the new temperature.

$$T\ (°C) = (\frac{1\ g\cdot°C}{4.18\ J})(175 \times 10^3\ J)(\frac{1}{1.00 \times 10^2\ g}) = 419°C$$

Since this temperature is impossible for liquid water, the water will boil first.

Calculate the amount of heat necessary to raise the temperature to 100°C

$$q\ (kJ) = (\frac{1\ kJ}{1000\ J})(\frac{4.18\ J}{1\ g\cdot°C})(10°C)(1.00 \times 10^2\ g) = 4.18\ kJ$$

175 - 4 = 171 kJ of energy available to boil the water. Calculate the amount of energy needed to boil all of the water

$$q\ (kJ) = (\frac{2.26\ kJ}{1\ g})(1.00 \times 10^2\ g) = 226\ kJ$$

Since there is not enough energy available to convert all of the water to steam, the remaining water will be boiling at 100°C after it has absorbed 175 kJ of energy.

13.89 The fire might have started from the vapors that were produced when the motor oil was heated. Raising the temperature increases the vapor pressure of a liquid. The vapors would burn when they encountered the gas flame. The investigator was not able to ignite a small volume of motor oil with a match because it was such a cold day that there was only a small amount of vapor in the absence of heating.

Chapter 14:
Solutions

14.1 Water is most common solvent.

14.3 Sodium ions (Na^+) and bicarbonate ions (HCO_3^-) are present.

14.5 A saturated NaCl solution is prepared by adding NaCl to water until so much has been added that not all of it dissolves. The undissolved solid sits in the bottom of the container. This solution could be made unsaturated by adding water to dissolve all the solid.

14.7 factors in solubility: the nature of intermolecular forces and temperature

14.9 Immiscible means do not dissolve in one another. Water and vegetable oil do not mix.

14.11 When KI dissolves, the ions are surrounded, or hydrated, by water molecules. The attractions between the ions and the water dipole are strong enough to counter the ion-ion attractions in the solid.

14.13 Ammonia gas is very soluble in water because it can form hydrogen bonds with water molecules. Nitrogen molecules are nonpolar and are not attracted to water molecules. Therefore, nitrogen is practically insoluble in water.

14.15 Hydrogen chloride, HCl(g), is very soluble in water due to ion-dipole attractions. Polar water molecules are attracted to the polar HCl molecules. The attractions are strong enough to ionize the HCl molecules, producing H^+(aq) and Cl^-(aq) ions surrounded by water molecules. Thus, even without hydrogen bonding, hydrogen chloride is able to dissolve in water.

14.17 36.4 g NaCl/100.0 g water

Find: solubility (g NaCl/100.0 g water) = ?
Given: 3.61 g NaCl undissolved from mixing 40.0 g NaCl with 100.0 g water
Known: solubility = g NaCl that will dissolve in 100.0 g water
Solution: The grams of NaCl dissolved = 40.0 g - 3.61 g = 36.4 g.
Therefore, the solubility is 36.4 g NaCl/100.0 g water.

14.19 The solubility of most solids in water increases when the temperature is increased.

14.21 A supersaturated solution is one that contains more solute than a saturated solution at that temperature. Supersaturated solutions are not stable.

14.23 When a supersaturated solution is allowed to stand for a period of time, the excess solute usually crystallizes until the remaining solution is saturated and therefore stable.

14.25 Since boiling decreases the solubility of gases, there is less oxygen in water that has been boiled and then cooled. Over time, oxygen diffuses back into the water, but if goldfish are placed in the cooled water right away, they will die for lack of oxygen.

14.27 A finely divided solid dissolves more rapidly than one composed of coarse particles because dissolving occurs at the surface. Finely divided solids have a larger surface area where water molecules can collide with solute particles and take them into solution.

14.29 5.0 g dextrose

Find: mass (g) dextrose = ?
Given: 5.0% dextrose (or $\frac{5.0 \text{ g dextrose}}{100 \text{ g solution}}$), 100. g solution
Solution: mass (g) dextrose = $(\frac{5.0 \text{ g dextrose}}{100 \text{ g solution}})(100. \text{ g solution}) = 5.0 \text{ g}$

14.31 (a) 3.8 g (b) 1.88 g (c) 0.14 g (d) 81.0

(a) Find: mass (g) NaCl = ?
Given: 5.0 % NaCl (or $\frac{5.0 \text{ g NaCl}}{100 \text{ g solution}}$), 75 g solution
Solution: mass (g) NaCl = $(\frac{5.0 \text{ g NaCl}}{100 \text{ g solution}})(75 \text{ g solution}) = 3.8 \text{ g NaCl}$

(b) Since this and the remaining parts are the same as part (a), only the solution will be shown.
Solution: mass (g) NaCl = $(\frac{1.50 \text{ g NaCl}}{100 \text{ g solution}})(125 \text{ g solution}) = 1.88 \text{ g NaCl}$
(c) Solution: mass (g) NaCl = $(\frac{0.25 \text{ g NaCl}}{100 \text{ g solution}})(55 \text{ g solution}) = 0.14 \text{ g NaCl}$
(d) Solution: mass (g) NaCl = $(\frac{12.0 \text{ g NaCl}}{100 \text{ g solution}})(675 \text{ g solution}) = 81.0 \text{ g NaCl}$

14.33 5.0 g NaOH and 95 g water

  Find: mass (g) water = ?, mass (g) NaOH = ?
  Given: 5.0% NaOH (or $\dfrac{5.0 \text{ g NaOH}}{100 \text{ g solution}}$), 100. g solution
  Solution: mass (g) NaOH = $(\dfrac{5.0 \text{ g NaOH}}{100 \text{ g solution}})(100.\text{ g solution})$ = 5.0 g NaOH

  Since 100. g solution contains 5.0 g NaOH, the balance of the solution must be water.

  100. g solution less 5.0 g NaOH = 95 g water.

14.35 0.50 g KMnO$_4$ and 24.5 g water

  Find: mass (g) KMnO$_4$ = ?, mass (g) water = ?
  Given: 2.0% KMnO$_4$ (or $\dfrac{2.0 \text{ g KMnO}_4}{100 \text{ g solution}}$), 25.0 g solution
  Solution: mass (g) KMnO$_4$ = $(\dfrac{2.0 \text{ g KMnO}_4}{100 \text{ g solution}})(25.0\text{ g solution})$ = 0.50 g KMnO$_4$

  Since 25.0 g solution contains 0.50 g KMnO$_4$, the balance of the solution must be water.

  25.0 g solution less 0.50 g KMnO$_4$ = 24.5 g water.

14.37 8.50 g HCl

  Find: mass (g) HCl = ?
  Given: 36.0 % HCl (or $\dfrac{36.0 \text{ g HCl}}{100 \text{ g solution}}$), $\dfrac{1.18 \text{ g solution}}{1 \text{ mL solution}}$, 20.0 mL solution
  Solution: mass (g) HCl = $(\dfrac{36.0 \text{ g HCl}}{100 \text{ g solution}})(\dfrac{1.18 \text{ g solution}}{1 \text{ mL solution}})(20.0\text{ mL solution})$ = 8.50 g HCl

14.39 0.25 g acetic acid

  Find: mass (g) acetic acid = ?

Given: 5.0% acetic acid (or $\frac{5.0 \text{ g acetic acid}}{100 \text{ g solution}}$), $\frac{1.006 \text{ g solution}}{1 \text{ mL vinegar}}$, 5.0 mL vinegar

Solution: mass (g) acetic acid = $(\frac{5.0 \text{ g acetic acid}}{100 \text{ g solution}})(\frac{1.006 \text{ g solution}}{1 \text{ mL vinegar}})(5.0 \text{ mL vinegar}) = 0.25$ g acetic acid

14.41 0.025 g NO_3^-

Find: mass (g) NO_3^- = ?
Given: 1.0 L water, 25 ppm (or $\frac{25 \text{ g } NO_3^-}{1 \times 10^6 \text{ g solution}}$), $\frac{1.0 \text{ g solution}}{1 \text{ mL solution}}$
Known: 1 L = 1000 mL
Solution: mass (g) NO_3^- = $(\frac{25 \text{ g } NO_3^-}{1 \times 10^6 \text{ g solution}})(\frac{1.0 \text{ g solution}}{1 \text{ mL solution}})(\frac{1000 \text{ mL solution}}{1 \text{ L solution}})(1.0 \text{ L solution})$

= 0.025 g NO_3^-

14.43 4.0 ppm F^-

Find: concentration (ppm) F^- = ?
Given: $\frac{2.0 \text{ mg } F^-}{0.50 \text{ L water}}$, $\frac{1.0 \text{ g water}}{1 \text{ mL water}}$
Known: 1 g = 1000 mg, 1 L = 1000 mL
Solution: concentration (ppm) F^- = $(\frac{1 \text{ g } F^-}{1000 \text{ mg } F^-})(\frac{2.0 \text{ mg } F^-}{0.50 \text{ L water}})(\frac{1 \text{ L water}}{1000 \text{ mL water}})(\frac{1 \text{ mL water}}{1.0 \text{ g water}})(10^6)$

= 4.0 ppm F^-

14.45 0.060 ppm Cr

Find: concentration (ppm) Cr = ?
Given: $\frac{0.060 \text{ mg Cr}}{1 \text{ L water}}$, $\frac{1.0 \text{ g water}}{1 \text{ mL water}}$

Known: 1 g = 1000 mg, 1 L = 1000 mL
Solution: concentration (ppm) Cr = $\left(\dfrac{1 \text{ g Cr}}{1000 \text{ mg Cr}}\right)\left(\dfrac{0.060 \text{ mg Cr}}{1 \text{ L water}}\right)\left(\dfrac{1 \text{ L water}}{1000 \text{ mL water}}\right)\left(\dfrac{1 \text{ mL water}}{1.0 \text{ g water}}\right)(10^6)$

= 0.060 ppm Cr

14.47 13.8 M H_2SO_4

Find: molarity (M) = ?

Given: 4.50 mol H_2SO_4, 325 mL (0.325 L) of solution

Solution: molarity (M) = $\dfrac{4.50 \text{ mol } H_2SO_4}{0.325 \text{ L solution}}$ = 13.8 M H_2SO_4

14.49 7.89 M HCl

Find: molarity (M) = ?

Given: 36.0 g HCl, 125 mL (0.125 L) solution

Known: MM-HCl = 36.5 g/mol HCl

Solution: First calculate moles of HCl = $\left(\dfrac{1 \text{ mol HCl}}{36.5 \text{ g HCl}}\right)(36.0 \text{ g HCl})$ = 0.986 mol HCl

molarity (M) = $\dfrac{0.986 \text{ mol HCl}}{0.125 \text{ L solution}}$ = 7.89 M HCl

14.51 5.00 M NaOH

Find: molarity (M) = ?

Given: 15.0 g NaOH, 75.0 mL (0.0750 L) solution

Known: MM-NaOH = 40.0 g/mol

Solution: First calculate moles of NaOH = $\left(\dfrac{1 \text{ mol NaOH}}{40.0 \text{ g NaOH}}\right)(15.0 \text{ g NaOH})$ = 0.375 mol NaOH

molarity (M) = $\dfrac{0.375 \text{ mol NaOH}}{0.0750 \text{ L solution}}$ = 5.00 M NaOH

14.53 (a) 0.900 M KCl   (b) 0.517 M Ca(NO$_3$)$_2$ (c) 0.0673 M BaCl$_2$ (d) 3.39 M KOH

(a) Find: molarity (M) KCl = ?
Given: 2.35 g KCl in 35.0 mL (0.0350 L) solution or $\dfrac{2.35 \text{ g KCl}}{0.0350 \text{ L solution}}$
Known: MM-KCl = 74.6 g/mol KCl
Solution: M = $\left(\dfrac{1 \text{ mol KCl}}{74.6 \text{ g KCl}}\right)\left(\dfrac{2.35 \text{ g KCl}}{0.0350 \text{ L solution}}\right)$ = 0.900 M KCl

(b) Find: molarity (M) Ca(NO$_3$)$_2$ = ?
Given: 10.6 g Ca(NO$_3$)$_2$ in 125 mL (0.125 L) solution or $\dfrac{10.6 \text{ g Ca(NO}_3)_2}{0.125 \text{ L solution}}$
Known: MM-Ca(NO$_3$)$_2$ = 164 g/mol Ca(NO$_3$)$_2$
Solution: M = $\left(\dfrac{1 \text{ mol Ca(NO}_3)_2}{164 \text{ g Ca(NO}_3)_2}\right)\left(\dfrac{10.6 \text{ g Ca(NO}_3)_2}{0.125 \text{ L solution}}\right)$ = 0.517 M Ca(NO$_3$)$_2$

(c) Find: molarity (M) BaCl$_2$ = ?
Given: 3.50 g BaCl$_2$ in 250. mL (0.250 L) solution or $\dfrac{3.50 \text{ g BaCl}_2}{0.250 \text{ L solution}}$
Known: MM-BaCl$_2$ = 208 g/mol BaCl$_2$
Solution: M = $\left(\dfrac{1 \text{ mol BaCl}_2}{208 \text{ g BaCl}_2}\right)\left(\dfrac{3.50 \text{ g BaCl}_2}{0.250 \text{ L solution}}\right)$ = 0.0673 M BaCl$_2$

(d) Find: molarity (M) KOH = ?
Given: 4.75 g KOH in 25.0 mL (0.0250 L) solution or $\dfrac{4.75 \text{ g KOH}}{0.0250 \text{ L solution}}$
Known: MM-KOH = 56.1 g/mol KOH
Solution: M = $\left(\dfrac{1 \text{ mol KOH}}{56.1 \text{ g KOH}}\right)\left(\dfrac{4.75 \text{ g KOH}}{0.0250 \text{ L solution}}\right)$ = 3.39 M KOH

14.55 (a) 50. g KI   (b) 21.3 g Na$_2$SO$_4$ (c) 23 g K$_3$PO$_4$ (d) 26.0 g BaCl$_2$

(a) Find: mass (g) KI = ?
Given: 100.0 mL (0.1000 L) solution, 3.0 M KI (or $\dfrac{3.0 \text{ mol KI}}{1 \text{ L solution}}$)
Known: MM-KI = 166 g/mol KI
Solution: mass (g) = $\left(\dfrac{166 \text{ g KI}}{1 \text{ mol KI}}\right)\left(\dfrac{3.0 \text{ mol KI}}{1 \text{ L solution}}\right)(0.1000 \text{ L solution})$ = 50. g KI

(b) Find: mass (g) Na$_2$SO$_4$ = ?
Given: 100.0 mL (0.1000 L) solution, 1.50 M Na$_2$SO$_4$ (or $\frac{1.50 \text{ mol Na}_2\text{SO}_4}{1 \text{ L solution}}$)

Known: MM-Na$_2$SO$_4$ = 142 g/mol Na$_2$SO$_4$
Solution: mass (g) = ($\frac{142 \text{ g Na}_2\text{SO}_4}{1 \text{ mol Na}_2\text{SO}_4}$)($\frac{1.50 \text{ mol Na}_2\text{SO}_4}{1 \text{ L solution}}$)(0.1000 L solution) = 21.3 g Na$_2$SO$_4$

(c) Find: mass (g) K$_3$PO$_4$ = ?
Given: 100.0 mL (0.1000 L) solution, 1.1 M K$_3$PO$_4$ (or $\frac{1.1 \text{ mol K}_3\text{PO}_4}{1 \text{ L solution}}$)

Known: MM-K$_3$PO$_4$ = 212 g/mol K$_3$PO$_4$
Solution: mass (g) = ($\frac{212 \text{ g K}_3\text{PO}_4}{1 \text{ mol K}_3\text{PO}_4}$)($\frac{1.1 \text{ mol K}_3\text{PO}_4}{1 \text{ L solution}}$)(0.1000 L solution) = 23 g K$_3$PO$_4$

(d) Find: mass (g) BaCl$_2$ = ?
Given: 100.0 mL (0.1000 L) solution, 1.25 M BaCl$_2$ (or $\frac{1.25 \text{ mol BaCl}_2}{1 \text{ L solution}}$)

Known: MM-BaCl$_2$ = 208 g/mol BaCl$_2$
Solution: mass (g) = ($\frac{208 \text{ g BaCl}_2}{1 \text{ mol BaCl}_2}$)($\frac{1.25 \text{ mol BaCl}_2}{1 \text{ L solution}}$)(0.1000 L solution) = 26.0 g BaCl$_2$

14.57 0.347 M C$_6$H$_{12}$O$_6$

Find: molarity (M) = ?
Given: 75.0 g C$_6$H$_{12}$O$_6$ in 1.20 L solution or $\frac{75.0 \text{ g C}_6\text{H}_{12}\text{O}_6}{1.20 \text{ L solution}}$

Known: MM-C$_6$H$_{12}$O$_6$ = 180. g/mol C$_6$H$_{12}$O$_6$
Solution: M = $\frac{1 \text{ mol C}_6\text{H}_{12}\text{O}_6}{180. \text{ g C}_6\text{H}_{12}\text{O}_6}$($\frac{75.0 \text{ g C}_6\text{H}_{12}\text{O}_6}{1.20 \text{ L solution}}$) = 0.347 M C$_6H_{12}O_6$

14.59 0.0488 mol sucrose

Find: number of moles sucrose = ?
Given: 65.0 mL (0.0650 L) solution, 0.750 M solution (or $\frac{0.750 \text{ mol sucrose}}{1 \text{ L solution}}$)

Solution: # moles sucrose = ($\frac{0.750 \text{ mol sucrose}}{1 \text{ L solution}}$)(0.0650 L solution) = 0.0488 mol sucrose

14.61 (a) 19.7 g Ca(NO$_3$)$_2$ (b) 20.0 g NaOH   (c) 7.31 g NaCl   (d) 567 g Na$_2$SO$_3$

(a) Find: mass (g) Ca(NO$_3$)$_2$ = ?
Given: 100. mL (0.100 L) solution, 1.20 M Ca(NO$_3$)$_2$ (or $\frac{1.20 \text{ mol Ca(NO}_3)_2}{1 \text{ L solution}}$)

Known: MM-Ca(NO$_3$)$_2$ = 164 g/mol Ca(NO$_3$)$_2$
Solution: mass (g) = ($\frac{164 \text{ g Ca(NO}_3)_2}{1 \text{ mol Ca(NO}_3)_2}$)($\frac{1.20 \text{ mol Ca(NO}_3)_2}{1 \text{ L solution}}$)(0.100 L solution) = 19.7 g Ca(NO$_3$)$_2$

(b) Find: mass (g) NaOH = ?
Given: 250. mL (0.250 L) solution, 2.00 M NaOH (or $\frac{2.00 \text{ mol NaOH}}{1 \text{ L solution}}$)

Known: MM-NaOH = 40.0 g/mol NaOH
Solution: mass (g) = ($\frac{40.0 \text{ g NaOH}}{1 \text{ mol NaOH}}$)($\frac{2.00 \text{ mol NaOH}}{1 \text{ L solution}}$)(0.250 L solution) = 20.0 g NaOH

(c) Find: mass (g) NaCl = ?
Given: 500. mL (0.500 L) solution, 0.250 M NaCl (or $\frac{0.250 \text{ mol NaCl}}{1 \text{ L solution}}$)

Known: MM-NaCl = 58.5 g/mol NaCl
Solution: mass (g) = ($\frac{58.5 \text{ g NaCl}}{1 \text{ mol NaCl}}$)($\frac{0.250 \text{ mol NaCl}}{1 \text{ L solution}}$)(0.500 L solution) = 7.31 g NaCl

(d) Find: mass (g) Na$_2$SO$_3$ = ?
Given: 1.50 L solution, 3.00 M Na$_2$SO$_3$ (or $\frac{3.00 \text{ mol Na}_2\text{SO}_3}{1 \text{ L solution}}$)

Known: MM-Na$_2$SO$_3$ = 126 g/mol Na$_2$SO$_3$
Solution: mass (g) = ($\frac{126 \text{ g Na}_2\text{SO}_3}{1 \text{ mol Na}_2\text{SO}_3}$)($\frac{3.00 \text{ mol Na}_2\text{SO}_3}{1 \text{ L solution}}$)(1.50 L solution) = 567 g Na$_2$SO$_3$

14.63 0.142 L (142 mL)

Find: V (L) = ?
Given: 0.0165 M HCl (or $\frac{0.0165 \text{ mol HCl}}{1 \text{ L solution}}$), 0.002 35 mol HCl
Solution: V (L) = ($\frac{1 \text{ L solution}}{0.0165 \text{ mol HCl}}$)(0.002 35 mol HCl) = 0.142 L (142 mL)

14.65 2.40 M H_2SO_4

Find: $M_d \left(\dfrac{\text{mol } H_2SO_4}{\text{L soln}}\right) = ?$

Given: 6.00 M H_2SO_4 $\left(\dfrac{6.00 \text{ mol } H_2SO_4}{\text{L soln}}\right)$, $V_c = 10.0$ mL (0.0100 L), $V_d = 25.0$ mL (0.0250 L)

Known: # moles H_2SO_4 does not change when solution is diluted

Solution: # moles $H_2SO_4 = \left(\dfrac{6.00 \text{ mol } H_2SO_4}{\text{L soln}}\right)(0.0100 \text{ L}) = 0.0600$ mol

$M_d \left(\dfrac{\text{mol } H_2SO_4}{\text{L soln}}\right) = \dfrac{0.0600 \text{ mol } H_2SO_4}{0.0250 \text{ L soln}} = 2.40$ M

14.67 0.60 M NH_3

Find: $M_d \left(\dfrac{\text{mol } NH_3}{\text{L soln}}\right) = ?$

Given: 15 M NH_3 $\left(\dfrac{15 \text{ mol } NH_3}{\text{L soln}}\right)$, $V_c = 10.0$ mL (0.0100 L), $V_d = 250.0$ mL (0.2500 L)

Known: # moles NH_3 does not change when solution is diluted

Solution: # moles $NH_3 = \left(\dfrac{15 \text{ mol } NH_3}{\text{L soln}}\right)(0.0100 \text{ L}) = 0.15$ mol

$M_d \left(\dfrac{\text{mol } NH_3}{\text{L soln}}\right) = \dfrac{0.15 \text{ mol } NH_3}{0.2500 \text{ L soln}} = 0.60$ M

14.69 2.00×10^{-3} L (2.00 mL)

Find: V_c (L) = ?

Given: $M_c = 0.125$ M $\left(\dfrac{0.125 \text{ mol } AgNO_3}{\text{L soln}}\right)$, $M_d = 0.0100$ M $\left(\dfrac{0.0100 \text{ mol } AgNO_3}{\text{L soln}}\right)$, $V_d = 0.0250$ L

Known: # moles $AgNO_3$ does not change when solution is diluted

Solution: # moles $AgNO_3 = \left(\dfrac{0.0100 \text{ mol } AgNO_3}{\text{L soln}}\right)(0.0250 \text{ L soln}) = 2.50 \times 10^{-4}$ mol

V_c (L) $= \left(\dfrac{1 \text{ L soln}}{0.125 \text{ mol } AgNO_3}\right)(2.50 \times 10^{-4} \text{ mol } AgNO_3) = 0.002\ 00$ L (2.00 mL)

14.71 0.25 L

Find: V_c (L) = ?
Given: M_c = 12 M ($\frac{12 \text{ mol HCl}}{\text{L soln}}$), M_d = 3.00 M ($\frac{3.00 \text{ mol HCl}}{\text{L soln}}$), V_d = 1.00 L

Known: # moles HCl does not change when solution is diluted
Solution: # moles HCl = ($\frac{3.00 \text{ mol HCl}}{\text{L soln}}$)(1.00 L soln) = 3.00 mol

V_c (L) = ($\frac{1 \text{ L soln}}{12 \text{ mol HCl}}$)(3.00 mol HCl) = 0.25 L

14.73 17.1 g CaSO$_4$

Outline: CaO(s) + H$_2$SO$_4$(aq) → H$_2$O(l) + CaSO$_4$(s)
         excess   18.6 mL   ? g

         6.75 M   MM = 136 g/mol

         ____ mol ($\frac{1 \text{ mol CaSO}_4}{1 \text{ mol H}_2\text{SO}_4}$) → ____ mol

moles H$_2$SO$_4$ = ($\frac{6.75 \text{ mol H}_2\text{SO}_4}{\text{L soln}}$)(0.0186 L) = 0.126 mol H$_2$SO$_4$

moles CaSO$_4$ = ($\frac{1 \text{ mol CaSO}_4}{1 \text{ mol H}_2\text{SO}_4}$)(0.126 mol H$_2SO_4$) = 0.126 mol CaSO$_4$

mass (g) CaSO$_4$ = ($\frac{136 \text{ g}}{\text{mol CaSO}_4}$)(0.126 mol CaSO$_4$) = 17.1 g CaSO$_4$

14.75 0.0622 g CaCl$_2$

Outline: Balance the equation and then complete the outline.

$$CaCl_2(aq) + 2\ AgNO_3(aq) \rightarrow 2\ AgCl(s) + Ca(NO_3)_2(aq)$$

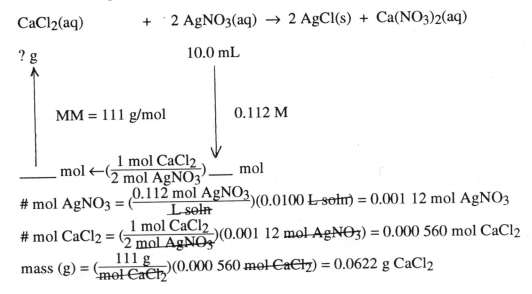

mol AgNO$_3$ = ($\dfrac{0.112 \text{ mol AgNO}_3}{\text{L soln}}$)(0.0100 L soln) = 0.001 12 mol AgNO$_3$

mol CaCl$_2$ = ($\dfrac{1 \text{ mol CaCl}_2}{2 \text{ mol AgNO}_3}$)(0.001 12 mol AgNO$_3$) = 0.000 560 mol CaCl$_2$

mass (g) = ($\dfrac{111 \text{ g}}{\text{mol CaCl}_2}$)(0.000 560 mol CaCl$_2$) = 0.0622 g CaCl$_2$

14.77 654 mL

Write the balanced equation and then complete the outline beneath it:

$$2\ NaOH(aq) + H_2SO_4(aq) \rightarrow Na_2SO_4(aq) + 2\ H_2O(l)$$

? mL
↑
0.765 M
|
____ mol ← ($\dfrac{2 \text{ mol NaOH}}{1 \text{ mol H}_2\text{SO}_4}$) 0.250 mol

mol NaOH = ($\dfrac{2 \text{ mol NaOH}}{1 \text{ mol H}_2\text{SO}_4}$)(0.250 mol H$_2SO_4$) = 0.500 mol NaOH

V (L) = ($\dfrac{1 \text{ L soln}}{0.765 \text{ mol NaOH}}$)(0.500 mol NaOH) = 0.654 L (654 mL)

Chapter 14: Solutions   169

14.79 0.252 L (252 mL)

Outline: $Fe_2O_3(s) + 4 H_3C_6H_5O_7(aq) \rightarrow 6 H^+(aq) + 2 Fe(C_6H_4O_7)_2{}^{3-}(aq) + 3 H_2O(l)$

 1.23 g ? L

 MM = 160. g/mol 0.122 M

 $\underline{}$ $(\dfrac{4 \text{ mol } H_3C_6H_5O_7}{1 \text{ mol } Fe_2O_3}) \rightarrow$ $\underline{}$ mol

mol Fe_2O_3 = $(\dfrac{1 \text{ mol } Fe_2O_3}{160. \text{ g}})(1.23 \text{ g})$ = 0.007 69 mol Fe_2O_3

mol $H_3C_6H_5O_7$ = $(\dfrac{4 \text{ mol } H_3C_6H_5O_7}{1 \text{ mol } Fe_2O_3})(0.007\,69 \text{ mol})$ = 0.0308 mol $H_3C_6H_5O_7$

V (L) = $(\dfrac{1 \text{ L soln}}{0.122 \text{ mol } H_3C_6H_5O_7})(0.0308 \text{ mol } H_3C_6H_5O_7)$ = 0.252 L (252 mL)

14.81 5.25 mL (5.25 x 10^{-3} L)

Write the balanced equation for the reaction and then complete the outline:

$H_3C_6H_5O_7(aq) + 3 NaOH(aq) \rightarrow Na_3C_6H_5O_7(aq) + 3 H_2O(l)$

 ? L 14.2 mL

 0.122 M 0.135 M

$\underline{}$ mol $\leftarrow (\dfrac{1 \text{ mol } H_3C_6H_5O_7}{3 \text{ mol } NaOH})$ $\underline{}$ mol

mol NaOH = $(\dfrac{0.135 \text{ mol}}{\text{L soln}})(0.0142 \text{ L})$ = 0.001 92 mol NaOH

mol H₃C₆H₅O₇ = ($\frac{1 \text{ mol H}_3\text{C}_6\text{H}_5\text{O}_7}{3 \text{ mol NaOH}}$)(0.001 92 mol NaOH) = 0.000 640 mol H₃C₆H₅O₇

V (L) = ($\frac{1 \text{ L soln}}{0.122 \text{ mol H}_3\text{C}_6\text{H}_5\text{O}_7}$)(0.000 640 mol H₃C₆H₅O₇) = 0.005 25 L (5.25 mL)

14.83 0.491 g NaHCO₃

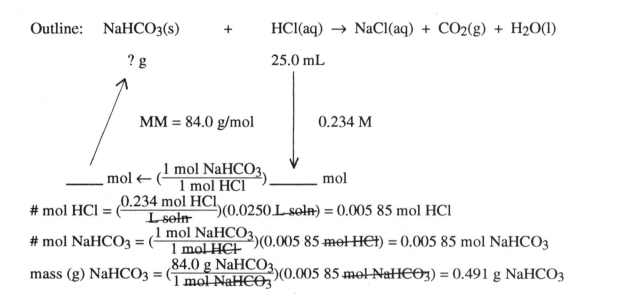

mol HCl = ($\frac{0.234 \text{ mol HCl}}{\text{L soln}}$)(0.0250 L soln) = 0.005 85 mol HCl

mol NaHCO₃ = ($\frac{1 \text{ mol NaHCO}_3}{1 \text{ mol HCl}}$)(0.005 85 mol HCl) = 0.005 85 mol NaHCO₃

mass (g) NaHCO₃ = ($\frac{84.0 \text{ g NaHCO}_3}{1 \text{ mol NaHCO}_3}$)(0.005 85 mol NaHCO₃) = 0.491 g NaHCO₃

14.85 102.9°C

Find: T_b (solution) = ?°C

Given: 175 g ethylene glycol (EG, C₂H₆O₂), 0.500 kg water

Known: MM-EG = 62.0 g EG/mol, T_b (solution) = 100°C + ΔT_b, K_b = 0.51°C/m

Solution: To calculate ΔT_b from ΔT_b = mK_b, you must first calculate m:

$m = \frac{\text{mol EG}}{0.500 \text{ kg water}}$

mol EG = ($\frac{1 \text{ mol EG}}{62.0 \text{ g EG}}$)(175 g EG) = 2.82 mol EG

$m = \frac{2.82 \text{ mol EG}}{0.500 \text{ kg water}} = 5.64 \text{ m}$

$\Delta T_b = (5.64\ m)(0.51°C/m) = 2.9°C$

$T_b = 100°C + \Delta T_b = 100 + 2.9 = 102.9°C$ (100 is an exact number)

14.87 100.45°C

Find: T_b (solution) = ?°C

Given: 225 g sucrose (sucrose, $C_{12}H_{22}O_{11}$), 750. g water

Known: MM-sucrose = 342 g sucrose/mol, T_b (solution) = 100°C + ΔT_b, $K_b = 0.51°C/m$

Solution: To calculate ΔT_b from $\Delta T_b = mK_b$, you must first calculate m:

$m = \dfrac{\text{mol sucrose}}{0.750\ \text{kg water}}$

\# mol sucrose = $(\dfrac{1\ \text{mol sucrose}}{342\ \text{g sucrose}})(225\ \text{g sucrose}) = 0.658$ mol sucrose

$m = \dfrac{0.658\ \text{mol sucrose}}{0.750\ \text{kg water}} = 0.877\ m$

$\Delta T_b = (0.877\ m)(0.51°C/m) = 0.45°C$

$T_b = 100°C + \Delta T_b = 100 + 0.45 = 100.45°C$ (100 is an exact number)

14.89 6.2 m

Find: molality (m) = ?

Given: $T_b = 103.2°C$

Known: $T_b = 100°C + \Delta T_b$, $\Delta T_b = mK_b$, $K_b = 0.51°C/m$

Solution: $m = \dfrac{\Delta T_b}{K_b}$

$\Delta T_b = T_b - 100°C = 103.2 - 100 = 3.2°C$

$m = \dfrac{\Delta T_b}{K_b} = \dfrac{3.2°C}{0.51°C/m} = 6.2\ m$

14.91 -119°C

Find: T_f (solution) = ?°C

Given: 2.50 g urea (CH_4N_2O), 50.0 g ethanol

Known: MM-urea = 60.0 g urea/mol, T_f (solution) = -117°C - ΔT_f, K_f = 1.99°C/m

Solution: To calculate ΔT_f from $\Delta T_f = mK_f$, you must first calculate m:

$$m = \frac{\text{mol urea}}{0.0500 \text{ kg ethanol}}$$

mol urea = $(\frac{1 \text{ mol urea}}{60.0 \text{ g urea}})(2.50 \text{ g urea}) = 0.0417$ mol urea

$$m = \frac{0.0417 \text{ mol urea}}{0.0500 \text{ kg ethanol}} = 0.834 \text{ m}$$

$\Delta T_f = (0.834 \text{ m})(1.99°C/m) = 1.66°C$

$T_f = -117°C + \Delta T_f = -117 - 1.66 = -119°C$

14.93 (a) 10.0 g urea in 100. g water has higher osmotic pressure

The more concentrated solution will have the higher osmotic pressure. Since there is an equal amount of the solvent in each case, it will be sufficient to calculate the number of moles of each solute. Known: MM-urea = 60.0 g/mol and MM-sucrose = 342 g/mol.

(a) # mol urea = $(\frac{1 \text{ mol urea}}{60.0 \text{ g urea}})(10.0 \text{ g urea}) = 0.167$ mol urea

(b) # mol sucrose = $(\frac{1 \text{ mol sucrose}}{342 \text{ g sucrose}})(50.0 \text{ g sucrose}) = 0.146$ mol

Therefore, solution (a) will have the higher osmotic pressure.

14.95 The Tyndall effect is the scattering of light by the particles in a colloid, which results in a foggy or cloudy appearance to the part of the solution through which the light is passing.

Chapter 14: Solutions

14.97 emulsions

Mayonnaise and homogenized milk are emulsions, a colloid formed by dispersing a liquid in a liquid.

14.99 0.838 M

Find: molarity (M) = ?
Given: 5.00% acetic acid (or $\frac{5.00 \text{ g HC}_2\text{H}_3\text{O}_2}{100 \text{ g soln}}$), density = $\frac{1.006 \text{ g soln}}{1 \text{ mL soln}}$
Known: MM-$HC_2H_3O_2$ = 60.0 g/mol
Solution: Since molarity is mol solute/L soln, assume 1 L of solution. To determine the moles of acetic acid, you need to know the mass of acetic acid. To determine the mass of acetic acid, you need to calculate the mass of 1 L of solution:

mass (g) soln = $(\frac{1.006 \text{ g soln}}{1 \text{ mL soln}})(\frac{1000 \text{ mL soln}}{1 \text{ L soln}})(1 \text{ L soln})$ = 1006 g soln

mass (g) $HC_2H_3O_2$ = $(\frac{5.00 \text{ g HC}_2\text{H}_3\text{O}_2}{100 \text{ g soln}})(1006 \text{ g soln})$ = 50.3 g $HC_2H_3O_2$

mol $HC_2H_3O_2$ = $(\frac{1 \text{ mol HC}_2\text{H}_3\text{O}_2}{60.0 \text{ g HC}_2\text{H}_3\text{O}_2})(50.3 \text{ g HC}_2\text{H}_3\text{O}_2)$ = 0.838 mol $HC_2H_3O_2$

molarity (M) = $\frac{0.838 \text{ mol HC}_2\text{H}_3\text{O}_2}{1 \text{ L soln}}$ = 0.838 M

14.101 3.39 M

Find: molarity (M) = ?
Given: 6.00% NH_3 (or $\frac{6.00 \text{ g NH}_3}{100 \text{ g soln}}$), density = $\frac{0.960 \text{ g soln}}{1 \text{ mL soln}}$
Known: MM-NH_3 = 17.0 g/mol

Solution: Since molarity is mol solute/L soln, assume 1 L of solution. To determine the moles of ammonia, you need to know the mass of ammonia. To determine the mass of ammonia, you need to calculate the mass of 1 L of solution:

$$\text{mass (g) soln} = \left(\frac{0.960 \text{ g soln}}{1 \text{ mL soln}}\right)\left(\frac{1000 \text{ mL soln}}{1 \text{ L soln}}\right)(1 \text{ L soln}) = 960. \text{ g soln}$$

$$\text{mass (g) NH}_3 = \left(\frac{6.00 \text{ g NH}_3}{100 \text{ g soln}}\right)(960. \text{ g soln}) = 57.6 \text{ g NH}_3$$

$$\text{\# mol NH}_3 = \left(\frac{1 \text{ mol NH}_3}{17.0 \text{ g NH}_3}\right)(57.6 \text{ g NH}_3) = 3.39 \text{ mol NH}_3$$

$$\text{molarity (M)} = \frac{3.39 \text{ mol NH}_3}{1 \text{ L soln}} = 3.39 \text{ M}$$

14.103   20.% HCl

Find: mass % = ?
Given: 6.0 M HCl $\left(\frac{6.0 \text{ mol HCl}}{\text{L soln}}\right)$, density = $\frac{1.10 \text{ g soln}}{1 \text{ mL soln}}$
Known: MM- HCl = 36.5 g/mol, mass % = $\left(\frac{\text{g HCl}}{\text{g soln}}\right)(100)$

Solution: Breaking this problem down into parts involves solving for mass of HCl and mass of solution. Use the density to calculate the mass of 1 L of solution. The mass of HCl in 1 L of solution can be calculated from the moles of HCl in 1 L of solution.

$$\text{mass (g) soln} = \left(\frac{1.10 \text{ g soln}}{1 \text{ mL soln}}\right)\left(\frac{1000 \text{ mL}}{1 \text{ L}}\right)(1 \text{ L}) = 1.10 \times 10^3 \text{ g soln}$$

$$\text{mass (g) HCl} = \left(\frac{36.5 \text{ g HCl}}{1 \text{ mol HCl}}\right)\left(\frac{6.0 \text{ mol HCl}}{\text{L soln}}\right)(1 \text{ L}) = 2.2 \times 10^2 \text{ g HCl}$$

$$\text{mass \%} = \left(\frac{2.2 \times 10^2 \text{ g HCl}}{1.10 \times 10^3 \text{ g soln}}\right)(100) = 20.\% \text{ HCl}$$

14.105   1.1 g/mL

Find: density (g/mL) = ?
Given: 34% HC$_2$H$_3$O$_2$ (or $\frac{34 \text{ g HC}_2\text{H}_3\text{O}_2}{100 \text{ g soln}}$), 6.0 M HC$_2H_3O_2$ (or $\frac{6.0 \text{ mol HC}_2\text{H}_3\text{O}_2}{\text{L soln}}$)

Known: MM-HC$_2$H$_3$O$_2$ = 60.0 g/mol HC$_2$H$_3$O$_2$

Solution: To calculate the density, it will be necessary to calculate the number of grams of solution in a given volume, say 1000 mL (1 L), of solution. This is not obvious. You can calculate the number of moles of HC$_2$H$_3$O$_2$ in 1 L of solution and then calculate the number of grams of HC$_2$H$_3$O$_2$ in this quantity of solution. Once you know the number of grams of HC$_2$H$_3$O$_2$, you can use the mass percent to calculate the mass of solution in this quantity. Now you can calculate the density. Remember all the calculations are based on one liter of solution:

$$\text{\# mol HC}_2\text{H}_3\text{O}_2 = \left(\frac{6.0 \text{ mol HC}_2\text{H}_3\text{O}_2}{\text{L soln}}\right)(1 \text{ L soln}) = 6.0 \text{ mol HC}_2\text{H}_3\text{O}_2$$

$$\text{mass (g) HC}_2\text{H}_3\text{O}_2 = \left(\frac{60.0 \text{ g HC}_2\text{H}_3\text{O}_2}{1 \text{ mol HC}_2\text{H}_3\text{O}_2}\right)(6.0 \text{ mol HC}_2\text{H}_3\text{O}_2) = 3.6 \times 10^2 \text{ g HC}_2\text{H}_3\text{O}_2$$

$$\text{mass (g) soln} = \left(\frac{100 \text{ g soln}}{34 \text{ g HC}_2\text{H}_3\text{O}_2}\right)(3.6 \times 10^2 \text{ g HC}_2\text{H}_3\text{O}_2) = 1.1 \times 10^3 \text{ g soln}$$

$$\text{density (g/mL)} = \frac{1.1 \times 10^3 \text{ g soln}}{1000 \text{ mL soln}} = 1.1 \text{ g/mL}$$

14.107 0.005 53 mol H$_2$O

This is a limiting reagent problem. Complete the outline, beginning with a balanced equation.

2 NaOH(aq) + H$_2$SO$_4$(aq) → Na$_2$SO$_4$(aq) + 2 H$_2$O(l)

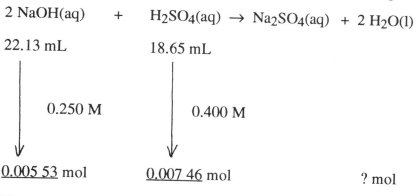

Hypothesis: NaOH is the limiting reactant.
Test: $(0.005\ 53 \text{ mol NaOH})\left(\frac{1 \text{ mol H}_2\text{SO}_4}{2 \text{ mol NaOH}}\right) = 0.002\ 67 \text{ mol H}_2\text{SO}_4$ needed

Conclusion: NaOH is the LR
Calculate # moles H$_2$O = $\left(\frac{2 \text{ mol H}_2\text{O}}{2 \text{ mol NaOH}}\right)(0.005\ 53 \text{ mol NaOH}) = 0.005\ 53 \text{ mol H}_2\text{O}$

Chapter 15:
Acids and Bases

15.1 Acids taste sour.

15.3 Vinegar tastes sour.

Vinegar is the ingredient traditionally used for the sour taste in sweet and sour recipes.

15.5 Bases taste bitter.

15.7 carbonates and bicarbonates

15.9 animal fats

15.11 (a) KOH and (c) Ba(OH)$_2$

Arrhenius bases dissolve in water to liberate hydroxide ions, OH$^-$. Only KOH and Ba(OH)$_2$ contain hydroxide ion.

15.13 (a) OH⁻, (b) H$_2$O, (c) Cl⁻, and (d) NH$_3$

All are Brønsted-Lowry bases because all are proton acceptors.

15.15 (a) Br⁻   (b) HSO$_4^-$ (c) CH$_3$NH⁻   (d) ClO$_3^-$

The conjugate base has one less proton (H$^+$) than the original acid.

15.17 (a) H⁻(aq) + H$_2$O(l) → H$_2$(g) + OH⁻(aq)

(b) NH$_2^-$(aq) + H$_2$O(l) → OH⁻(aq) + NH$_3$(aq)

(c) O^{2-}(aq) + H$_2$O(l) → OH⁻(aq) + OH⁻(aq)   [O^{2-}(aq) + H$_2$O(l) → 2 OH⁻(aq)]

(d) NH$_3$(aq) + H$_2$O(l) → OH⁻(aq) + NH$_4^+$(aq)

(e) CO$_3^{2-}$(aq) + H$_2$O(l) → OH⁻(aq) + HCO$_3^-$(aq)

(f) CN⁻(aq) + H$_2$O(l) → OH⁻(aq) + HCN(aq)

15.19 (a), (c), and (d) are amphoteric. Conjugate acids: (a) H$_2$O, (c) H$_2$S, (d) H$_3$PO$_4$; conjugate bases: (a) O^{2-}, (c) S^{2-}, (d) HPO$_4^{2-}$

To be an acid, a substance must have a hydrogen it can donate. All four substances have hydrogen. To behave as a base, a substance must either be an anion or have an atom with a lone pair of electrons that can accept a hydrogen ion. C$_2$H$_6$ does not qualify as a base.

15.21 0.126 M HCl

Find: $M \left(\dfrac{\text{mol HCl}}{\text{L soln}} \right) = ?$

Given: $\dfrac{0.133 \text{ mol NaOH}}{\text{L soln}}$, 20.0 mL (0.0200 L) HCl soln, 18.9 mL (0.0189 L) NaOH soln

Known: $HCl(aq) + NaOH(aq) \rightarrow NaCl(aq) + H_2O(l)$

Solution: From the equation, we have $\dfrac{1 \text{ mol HCl}}{1 \text{ mol NaOH}}$

number of moles HCl = $(\dfrac{1 \text{ mol HCl}}{1 \text{ mol NaOH}})$(? mol NaOH)

number of moles NaOH = $(\dfrac{0.133 \text{ mol NaOH}}{\text{L soln}})$(0.0189 L̶ ̶s̶o̶l̶n̶) = 0.002 51 mol NaOH

∴ number of moles HCl = $(\dfrac{1 \text{ mol HCl}}{1 \text{ mol NaOH}})$(0.002 51 m̶o̶l̶ ̶N̶a̶O̶H̶) = 0.002 51 mol HCl

$M (\dfrac{\text{mol HCl}}{\text{L soln}}) = \dfrac{0.002\ 51 \text{ mol HCl}}{0.0200 \text{ L soln}} = 0.126 \text{ M HCl}$

15.23 0.160 M NaOH

Find: $M (\dfrac{\text{mol NaOH}}{\text{L soln}}) = ?$

Given: 0.235 g $H_2C_2O_4 \cdot 2H_2O$, 23.3 mL (0.0233 L) NaOH soln

Known: $H_2C_2O_4(aq) + 2\ NaOH(aq) \rightarrow Na_2C_2O_4(aq) + 2\ H_2O(l)$

MM-$H_2C_2O_4 \cdot 2H_2O$ = 126 g/mol

$\dfrac{1 \text{ mol } H_2C_2O_4}{1 \text{ mol } H_2C_2O_4 \cdot 2H_2O}$

Solution: From the equation, we have $\dfrac{1 \text{ mol } H_2C_2O_4}{2 \text{ mol NaOH}}$

number of moles NaOH = $(\dfrac{2 \text{ mol NaOH}}{1 \text{ mol } H_2C_2O_4})$(? mol $H_2C_2O_4$)

number of moles $H_2C_2O_4$ = $(\dfrac{1 \text{ mol } H_2C_2O_4}{126 \text{ g̶ ̶H̶}_2\text{C̶}_2\text{O̶}_4 \text{ ̶·̶ ̶2̶H̶}_2\text{O̶}})$(0.235 g̶ ̶H̶_2C̶_2O̶_4·̶2̶H̶_2O̶) = 0.001 87 mol $H_2C_2O_4$

∴ number of moles NaOH = $(\dfrac{2 \text{ mol NaOH}}{1 \text{ m̶o̶l̶ ̶H̶}_2\text{C̶}_2\text{O̶}_4})$(0.001 87 m̶o̶l̶ ̶H̶_2C̶_2O̶_4) = 0.003 74 mol NaOH

$M (\dfrac{\text{mol NaOH}}{\text{L soln}}) = \dfrac{0.003\ 74 \text{ mol NaOH}}{0.0233 \text{ L soln}} = 0.160 \text{ M NaOH}$

15.25 An acid anhydride is a compound that reacts with water to form an acid. SO_2 is an example involving an element in Group 6A: $SO_2(g) + H_2O(l) \rightarrow H_2SO_3(aq)$.

15.27 $H_2SO_4(aq)$

The acid is obtained by combining SO_3 and H_2O: $SO_3(g) + H_2O(l) \rightarrow H_2SO_4(aq)$.

15.29 $Cl_2O(g) + H_2O(l) \rightarrow 2\ HClO(aq)$. The oxidation number of chlorine is +1 in each compound.

15.31 (a) NaOH   (b) $Ba(OH)_2$ (d) $Cu(OH)_2$

15.33 $ZnO(s) + 2\ HCl(aq) \rightarrow ZnCl_2(aq) + H_2O(l)$

15.35 $3\ H_2SO_4(aq) + Al_2O_3(s) \rightarrow Al_2(SO_4)_3(aq) + 3\ H_2O(l)$

15.37 A weak acid is one that is only slightly ionized in dilute solution. The notation using unequal arrows in the equation below indicates that most of the acetic acid molecules are not ionized.

$HC_2H_3O_2(aq) + H_2O(l) \rightleftharpoons C_2H_3O_2^-(aq) + H_3O^+(aq)$

15.39 F^-

15.41 $OH^- > NH_3 > F^- > Cl^- > ClO_4^-$

15.43 neutral solution: $[H_3O^+] = [OH^-] = 1.0 \times 10^{-7}$ mol/L

15.45 neutral solution: pH = pOH = 7

15.47 pH = 1.0

$pH = -\log[H_3O^+] = -\log(0.1) = -\log(1 \times 10^{-1}) = 1.0$

15.49 Sourness depends on the acidity of the solution. The pH of orange juice is higher (3.7) than that of lemon juice (2.3). The higher the pH, the less acidic the solution. Therefore, orange juice is less acidic and less sour than lemon juice. Additionally, orange juice may have more natural sugar than lemon juice, with the sweetness masking some of the sour taste.

15.51 (a) pOH = 9.0, [OH$^-$] = 1 x 10^{-9} M (c) pOH = 11.0, [OH$^-$] = 1 x 10^{-11} M
 (b) pOH = 6.0, [OH$^-$] = 1 x 10^{-6} M (d) pOH = 2.0, [OH$^-$] = 1 x 10^{-2} M

Since pH + pOH = 14.0, then pOH = 14.0 - pH. [OH$^-$] = antilog(-pOH) = 10^{-pOH}. For integer values of pOH, as in this problem, the antilog is obtained by substituting the negative value of pOH for the exponent of ten. (a) pOH = 14.0 - 5.0 = 9.0, [OH$^-$] = antilog(-9.0) = 10^{-9} M; (b) pOH = 14.0 - 8.0 = 6.0, [OH$^-$] = antilog(-6.0) = 10^{-6} M; (c) pOH = 14.0 - 3.0 = 11.0, [OH$^-$] = antilog(-11.0) = 10^{-11} M; (d) pOH = 14.0 - 12.0 = 2.0, [OH$^-$] = antilog(-2.0) = 10^{-2} M

15.53 pH meter

15.55 most common type of buffer solution: solution consisting of a mixture of a weak acid and its conjugate base

15.57 $H_2PO_4^-/HPO_4^{2-}$ (acid/conjugate base)

15.59 An acetic acid buffer resists a change in pH when a strong acid is added by converting some of the acetate ion into its conjugate acid, acetic acid. In the process, the hydronium ions from the strong acid are consumed, and there is very little change in pH.

$$C_2H_3O_2^-(aq) + H_3O^+(aq) \rightarrow HC_2H_3O_2(aq) + H_2O(l)$$

15.61 When strong acid is added to the HCO_3^-/CO_3^{2-} buffer, the carbonate ion reacts with the hydronium ion to form the bicarbonate ion. Thus, there is no significant increase in the concentration of hydronium ion, and the pH is unchanged.

$$CO_3^{2-}(aq) + H_3O^+(aq) \rightarrow HCO_3^-(aq) + H_2O(l)$$

15.63 0.156 M HCO_2H

Find: $M(\frac{mol\ HCO_2H}{L\ soln}) = ?$

Given: 20.0 mL (0.0200 L) HCO_2H, 23.4 mL (0.0234 L) KOH, $\frac{0.133\ mol\ KOH}{L\ soln}$

Known: $HCO_2H(aq) + KOH(aq) \rightarrow KHCO_2(aq) + H_2O(l)$

Solution: From the equation, we have $\frac{1\ mol\ HCO_2H}{1\ mol\ KOH}$

number of moles $HCO_2H = (\frac{1\ mol\ HCO_2H}{1\ mol\ KOH})$(? mol KOH)

number of moles $KOH = (\frac{0.133\ mol\ KOH}{L\ soln})(0.0234\ L\ soln) = 0.003\ 11$ mol KOH

∴ number of moles $HCO_2H = (\frac{1\ mol\ HCO_2H}{1\ mol\ KOH})(0.003\ 11\ mol\ KOH)$

= 0.003 11 mol HCO_2H

$M(\frac{mol\ HCO_2H}{L\ soln}) = \frac{0.003\ 11\ mol\ HCO_2H}{0.0200\ L\ soln} = 0.156\ M\ HCO_2H$

15.65 first step: $H_2SO_4(aq) + H_2O(l) \rightarrow HSO_4^-(aq) + H_3O^+(aq)$

second step: $HSO_4^-(aq) + H_2O(l) \rightarrow SO_4^{2-}(aq) + H_3O^+(aq)$

HSO_4^- is a weaker acid than H_2SO_4 because the negatively charged ion holds onto its positively charged proton more strongly than does the neutral molecule.

15.67 pH = 2.82

Since HBr is a strong acid, it ionizes completely. Therefore the concentration of hydronium ion is $[H_3O^+] = 0.0015$ M. pH = $- \log[H_3O^+]$ = $- \log(0.0015) = 2.82$ (two significant figures).

15.69 (a) base anhydride, $Zn(OH)_2$ (c) acid anhydride, H_3AsO_4

(b) base anhydride, $Al(OH)_3$ (d) acid anhydride, H_2SeO_3

Oxides of metals are base anhydrides. Oxides of nonmetals are acid anhydrides. Water is added to the formulas to make an acid or base in which the element has the same oxidation number.

15.71 4×10^{-3} M

pOH = 14.0 - pH = 14.0 - 11.6 = 2.4. $[OH^-]$ = antilog(-pOH) = antilog(-2.4) = $10^{-2.4}$ = 4×10^{-3} M (one significant figure because there is only one decimal place in the pH value.)

Chapter 16:
Reaction Rates and Chemical Equilibrium

16.1 Reaction kinetics means the study of reaction rates, or how fast reactions occur.

16.3 The basic assumption of the collision theory of reactions is that reacting particles must collide for a reaction to occur.

16.5 Nothing happens to the reactants in an ineffective collision—they bounce apart with no change.

16.7 Large complex molecules tend to react more slowly in a chemical reaction than small molecules because they are less likely to be oriented properly when they collide.

16.9 The activation energy of a chemical reaction is the minimum amount of kinetic energy required for colliding molecules to react.

16.11 For an exothermic reaction, the energy state of products is lower than the energy state of reactants.

16.13

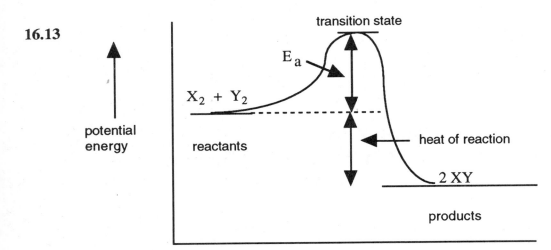

16.15 The larger the value of E_a, the slower the rate of a chemical reaction.

16.17 Raising the temperature increases the rate of a chemical reaction.

16.19 Raising the temperature increases the rate of a reaction by increasing the number of molecules that have kinetic energies greater than E_a. Therefore, more collisions will be effective at higher temperatures. Additionally, as the particles move faster, more collisions will occur in a given time.

16.21 A catalyst is a substance that speeds up the rate of a reaction without being consumed by the reaction.

16.23 Adding a catalyst provides an additional pathway that has a lower activation energy. The reaction will primarily occur by this alternate pathway, since a larger number of particles will have sufficient energy to react by the catalyzed pathway.

16.25 KI serves as a catalyst for the decomposition of aqueous hydrogen peroxide.

16.27 Increasing the concentrations of reactants increases the number of collisions that occur in a given period of time. If there are more frequent collisions, there will be more frequent effective collisions. As a result, the rate of the reaction will increase.

16.29 Chemical equilibrium is the state in which the rate of the forward reaction equals the rate of the reverse reaction.

16.31 The concentrations of reactants and products remains unchanged at equilibrium.

16.33 Equilibrium is achieved rapidly at 250°C because the rates of reactions increase with increasing temperature. Since equilibrium is the condition in which the rates of forward and reverse reactions are equal, the faster the reactions occur, the sooner will equilibrium be reached.

16.35 $K_{eq} = \dfrac{[NO]^4[H_2O]^6}{[NH_3]^4[O_2]^5}$

The concentrations of products are written in the numerator within brackets, with each concentration raised to the power corresponding to its coefficient in the balanced equation. The concentrations of reactants are placed in the denominator in the same fashion as the products in the numerator.

16.37 1.67×10^3

First write a balanced chemical equation for the reaction.

$$SO_2(g) + Cl_2(g) \rightleftharpoons SO_2Cl_2(g)$$

Write the K_{eq} expression, substitute values, and calculate the value of K_{eq}:

$$K_{eq} = \frac{[SO_2Cl_2]}{[SO_2][Cl_2]} = \frac{(0.952)}{(0.0239)(0.0239)} = 1.67 \times 10^3$$

16.39 $K_{eq} = \dfrac{[CO_2]^2}{[CO]^2[O_2]} = 5.0 \times 10^3$. The formation of CO_2 is favored. The large value of K_{eq} indicates a large concentration of product (CO_2) and a small concentration of reactants at equilibrium.

16.41 (a) $N_2(g) + 2 O_2(g) \rightleftharpoons 2 NO_2(g)$ (b) $K_{eq} = \dfrac{[NO_2]^2}{[N_2][O_2]^2} = \dfrac{1}{0.010} = 1.0 \times 10^2$

(c) The reaction is favored in the direction represented in this problem, as indicated by the value of K_{eq} greater than 1.

16.43 Le Chatelier's principle states that if a reaction at equilibrium is changed, the reaction will adjust to counteract the change and establish a new equilibrium.

16.45 (a) decreases   (b) decreases

Raising the temperature of an exothermic reaction will result in a shift to the left (the endothermic direction, the direction in which heat is absorbed). (a) Since NOBr is a product, its equilibrium concentration will decrease. (b) The value of K_{eq} will decrease.

16.47 Adding chlorine, a product, to the equilibrium mixture in problem 46 will cause the reaction to shift to the left. Reaction to the left is the exothermic direction. Therefore, heat will be released when chlorine is added.

16.49 (a) shift right (b) shift right (c) shift right (d) shift right

(a) removing product causes a shift to increase product, (b) decreasing temperature causes a shift in the exothermic direction which is forward, (c) adding reactant causes a shift away from the reactant, (d) decreasing volume causes a shift in the direction that reduces the number of gas molecules.

16.51 Adding NaOH causes the equilibrium to shift away from the reactant side, increasing the concentration of CrO_4^{2-}(aq), which is yellow. Therefore, the solution will turn yellow.

16.53 (a) increases (b) increases (c) no change

The presence of a catalyst increases the rates of both the forward and the reverse directions for a reaction. Both directions are sped up. Equilibrium is the state at which both rates are equal. Since *both* rates speed up, there is no change in the position of equilibrium—the reaction just gets to equilibrium faster in the presence of a catalyst.

16.55 (a) $K_b = \dfrac{[NH_4^+][OH^-]}{[NH_3]}$ (b) $K_b = \dfrac{[C_2H_5NH_3^+][OH^-]}{[C_2H_5NH_2]}$

16.57 0.1% ionized

Find: % ionization = ?
Given: 0.10 M HA, pH = 4.0
Known: HA(aq) $\rightleftharpoons$ H$^+$(aq) + A$^-$(aq)

Solution: When pH = 4.0, [H$^+$] = 1 x 10^{-4} M, or 0.0001 M. The chemical equation for the ionization of HA shows [H$^+$] = [A$^-$], and only 0.0001 mol/L of the 0.10 mol/L HA is ionized. Calculate the percent ionization:

$$\% \text{ ionization} = \frac{(0.0001)}{(0.10)} \times 100 = 0.1\%$$

16.59 $K_a = 2 \times 10^{-6}$

Find: $K_a = ?$

Given: 0.20 M HX, pH = 3.2

Known: [H$^+$] = antilog(-3.2) = 0.0006 M, [X$^-$] = [H$^+$], $K_a = \frac{[H^+][X^-]}{[HX]}$

Solution: $K_a = \frac{[H^+][X^-]}{[HX]} = \frac{(0.0006)(0.0006)}{(0.20)} = 2 \times 10^{-6}$

16.61 0.03%

Find: % ionization = ?

Given: 0.10 M B, pH = 9.4

Known: B(aq) + H$_2$O(l) ⇌ BH$^+$(aq) + OH$^-$(aq)

Solution: When pH = 9.4, [H$^+$] = 4 x 10^{-10} M, and [OH$^-$] = 3 x 10^{-5} M or 0.000 03 M (1 significant figure). The chemical equation for the ionization of B shows [BH$^+$] = [OH$^-$], and only 0.000 03 mol/L of the 0.10 mol/L B is ionized. Calculate the percent ionization:

$$\% \text{ ionization} = \frac{(0.000\ 03)}{(0.10)} \times 100 = 0.03\%$$

16.63 pH = 2.41

Chapter 16: Reaction Rates and Chemical Equilibrium   191

Find: pH = ?

Given: 0.11 M lactic acid (HL), $K_a = 1.4 \times 10^{-4}$

Known: pH = -log[H$^+$], $K_a = \dfrac{[H^+][L^-]}{[HL]}$

$HL(aq) \rightleftharpoons H^+(aq) + L^-(aq)$

Solution: Let [H$^+$] = [L$^-$] = x

Because K_a is about 10^{-4}, the percent ionization of the acid is very small, and the concentration at equilibrium of the acid, [HL], $\cong$ 0.11 M.

$K_a = 1.4 \times 10^{-4} = \dfrac{[H^+][L^-]}{[HL]} = \dfrac{(x)(x)}{(0.11)} = \dfrac{(x^2)}{(0.11)}$

$x^2 = (0.11)(1.4 \times 10^{-4}) = 1.5 \times 10^{-5}$

$x = 3.9 \times 10^{-3} = [H^+]$

pH = -log(3.9 × 10^{-3}) = 2.41

16.65 (a) $C_4H_{10}(g) \underset{}{\overset{catalyst}{\rightleftharpoons}} C_4H_8(g) + H_2(g)$; $K_{eq} = \dfrac{[C_4H_8][H_2]}{[C_4H_{10}]}$ (b) $K_{eq} = 14$

$K_{eq} = \dfrac{[C_4H_8][H_2]}{[C_4H_{10}]} = \dfrac{(4.5)(4.5)}{(1.5)} = 14$

16.67 $K_{eq} = 3.2 \times 10^7$

$K_{eq} = \dfrac{[NO_2]^2}{[N_2][O_2]^2} = \dfrac{(4.5 \times 10^{-2})^2}{(1.2 \times 10^{-3})(2.3 \times 10^{-4})^2} = 3.2 \times 10^7$

16.69 (a) shift right   (b) shift left   (c) shift right   (d) shift right

For parts (a), (b), and (c), the reaction will shift away from the substance added. For part (d), the reaction will shift right to produce fewer gas molecules and occupy less volume.

16.71 (a) $N_2H_4(aq) + H_2O(l) \rightleftharpoons N_2H_5^+(aq) + OH^-(aq)$

(b) $K_{eq} = \dfrac{[N_2H_5^+][OH^-]}{[N_2H_4]}$ (c) 0.008%

(c) Find: % ionization = ?

Given: 0.25 M hydrazine, pH = 9.3

Known: $[H^+]$ = antilog(-9.3) = 5 x 10^{-10} M

Solution: $[OH^-]$ = 2 x 10^{-5} M

$\%\text{ ionization} = \dfrac{(2 \times 10^{-5})}{(0.25)} \times 100 = 0.008\%$

Chapter 17:
Oxidation-Reduction Reactions

17.1 Oxidation means loss of electrons. Metals tend to undergo oxidation readily.

17.3 An oxidizing agent causes a substance to be oxidized by accepting its electrons. The term originated with oxygen, a common oxidizing agent.

17.5 A reducing agent causes a substance to be reduced by giving it electrons. Metals are the best reducing agents.

17.7 (a) Mn^{2+} and Fe^{3+} (b) $NO(g)$ and Sn^{2+} (c) no reaction

(a) Since MnO_4^- is an oxidizing agent and Fe^{2+} is a reducing agent, reaction is likely to occur. The products were found in Table 17.1. (b) Sn is a reducing agent and NO_3^- is an oxidizing agent. Therefore, reaction is likely to occur, giving the products listed in Table 17.1. (c) Since $Cl_2(g)$ and Cu^{2+} are both oxidizing agents, they will not react with each other.

17.9 oxidizing agent is reduced: $Cl_2(g)$;   reducing agent is oxidized: Cu

17.11 $Mg(s) + Cu^{2+}(aq) \rightarrow Mg^{2+}(aq) + Cu(s)$

17.13 oxidizing agent is reduced: $Cu^{2+}(aq)$;   reducing agent is oxidized: $Mg(s)$

17.15 Mn is +7 in MnO_4^- and N is +5 in NO_3^-

17.17 element oxidized: Ag; element reduced: Mn; oxidizing agent: $MnO_4^-(aq)$; reducing agent: $Ag(s)$

17.19 element oxidized: Cu; element reduced: N; oxidizing agent: $NO_3^-(aq)$; reducing agent: $Cu(s)$

17.21 element oxidized: Cu; element reduced: Cr; oxidizing agent: $Cr_2O_7^{2-}(aq)$; reducing agent: $Cu(s)$

17.23 (a) $3\,CuS(s) + 2\,NO_3^-(aq) + 8\,H^+(aq) \rightarrow 3\,Cu^{2+}(aq) + 3\,S(s) + 2\,NO(g) + 4\,H_2O(l)$

    (b) $ClO^-(aq) + 2\,I^-(aq) + 2\,H^+(aq) \rightarrow Cl^-(aq) + I_2(aq) + H_2O(l)$

    (c) $6\,Zn(s) + As_2O_3(s) + 12\,H^+(aq) \rightarrow 6\,Zn^{2+}(aq) + 2\,AsH_3(g) + 3\,H_2O(l)$

    (a) Step 1: Separate into half-reactions:

$$CuS \rightarrow Cu^{2+} + S$$

$$NO_3^- \rightarrow NO(g)$$

        Step 2: Balance the equations for the half-reactions:

$$CuS \rightarrow Cu^{2+} + S + 2\,e^-$$

$$NO_3^- + 4\,H^+ + 3\,e^- \rightarrow NO + 2\,H_2O$$

        Step 3: $3(CuS \rightarrow Cu^{2+} + S + 2\,e^-) = 3\,CuS \rightarrow 3\,Cu^{2+} + 3\,S + 6\,e^-$

$$\underline{2(NO_3^- + 4\,H^+ + 3\,e^- \rightarrow NO + 2\,H_2O) = 2\,NO_3^- + 8\,H^+ + 6\,e^- \rightarrow 2\,NO + 4\,H_2O}$$

$$3\,CuS + 2\,NO_3^- + 8\,H^+ \rightarrow 3\,Cu^{2+} + 3\,S + 2\,NO + 4\,H_2O$$

Step 4: Check the equation.

Step 5: Add symbols to indicate whether the substances are solid, liquid, gas, or in aqueous solution.

(b) Step 1: ClO⁻ → Cl⁻

I⁻ → I₂

Step 2: ClO⁻ + 2 H⁺ + 2 e⁻ → Cl⁻ + H₂O

2 I⁻ → I₂ + 2 e⁻

Step 3: ClO⁻ + 2 I⁻ + 2 H⁺ → Cl⁻ + I₂ + H₂O

Step 4: Check.

Step 5: Add symbols for physical states.

(c) Step 1: Zn → Zn²⁺

As₂O₃ → AsH₃

Step 2: Zn → Zn²⁺ + 2 e⁻

As₂O₃ + 12 H⁺ + 12 e⁻ → 2 AsH₃ + 3 H₂O

Step 3: 6(Zn → Zn²⁺ + 2 e⁻) = 6 Zn → 6 Zn²⁺ + 12 e⁻

As₂O₃ + 12 H⁺ + 12 e⁻ → 2 AsH₃ + 3 H₂O

6 Zn + As₂O₃ + 12 H⁺ → 6 Zn²⁺ + 2 AsH₃ + 3 H₂O

Steps 4 and 5: Check. Then add symbols for physical states.

17.25 (a) $S_2O_3^{2-}(aq) + 4\ Cl_2(g) + 5\ H_2O(l) \rightarrow 2\ HSO_4^-(aq) + 8\ Cl^-(aq) + 8\ H^+(aq)$

(b) $3\ AsH_3(g) + 4\ ClO_3^-(aq) \rightarrow 3\ H_3AsO_4(aq) + 4\ Cl^-(aq)$

(c) $3\ P_4(s) + 20\ NO_3^-(aq) + 8\ H_2O(l) \rightarrow 12\ HPO_4^{2-}(aq) + 20\ NO(g) + 4\ H^+(aq)$

(a) Step 1: Separate into half-reactions:

$S_2O_3^{2-} \rightarrow HSO_4^-$

$Cl_2 \rightarrow Cl^-$

Step 2: Balance each half-reaction:

$S_2O_3^{2-} + 5 H_2O \rightarrow 2 HSO_4^- + 8 H^+ + 8 e^-$

$Cl_2 + 2 e^- \rightarrow 2 Cl^-$

Step 3: Combine half-reactions with cancellation of electrons:

$S_2O_3^{2-} + 5 H_2O \rightarrow 2 HSO_4^- + 8 H^+ + 8 e^-$

$\underline{4(Cl_2 + 2 e^- \rightarrow 2 Cl^-)}$

$S_2O_3^{2-} + 4 Cl_2 + 5 H_2O \rightarrow 2 HSO_4^- + 8 Cl^- + 8 H^+$

Steps 4 & 5: Check and add physical states.

(b) Step 1: $AsH_3 \rightarrow H_3AsO_4$

$ClO_3^- \rightarrow Cl^-$

Step 2: $AsH_3 + 4 H_2O \rightarrow H_3AsO_4 + 8 H^+ + 8 e^-$

$ClO_3^- + 6 H^+ + 6 e^- \rightarrow Cl^- + 3 H_2O$

Step 3: $3(AsH_3 + 4 H_2O \rightarrow H_3AsO_4 + 8 H^+ + 8 e^-)$

$\underline{4(ClO_3^- + 6 H^+ + 6 e^- \rightarrow Cl^- + 3 H_2O)}$

$3 AsH_3 + 4 ClO_3^- \rightarrow 3 H_3AsO_4 + 4 Cl^-$

Steps 4 & 5: Check and add physical states.

(c) Step 1: $P_4 \rightarrow HPO_4^{2-}$

$NO_3^- \rightarrow NO$

Step 2: $P_4 + 16 H_2O \rightarrow 4 HPO_4^{2-} + 28 H^+ + 20 e^-$

$NO_3^- + 4 H^+ + 3 e^- \rightarrow NO + 2 H_2O$

Step 3: $3(P_4 + 16 H_2O \rightarrow 4 HPO_4^{2-} + 28 H^+ + 20 e^-)$

$\underline{20(NO_3^- + 4 H^+ + 3 e^- \rightarrow NO + 2 H_2O)}$

$3 P_4 + 48 H_2O + 20 NO_3^- + 80 H^+ \rightarrow 12 HPO_4^{2-} + 84 H^+ + 20 NO + 40 H_2O$

$3 P_4 + 20 NO_3^- + 8 H_2O \rightarrow 12 HPO_4^{2-} + 20 NO + 4 H^+$

Steps 4 & 5: Check and add symbols for physical states.

17.27 (a) $N_2(g) + O_2(g) \rightarrow 2\,NO(g)$ and $2\,NO(g) + O_2(g) \rightarrow 2\,NO_2(g)$

(b) oxidizing agent in each reaction is O_2; reducing agents are N_2 and NO

17.29 (a) $2\,MnO(s) + 5\,PbO_2(s) + 8\,H^+(aq) \rightarrow 2\,MnO_4^-(aq) + 5\,Pb^{2+}(aq) + 4\,H_2O(l)$

(b) $3\,SO_2(g) + 2\,NO_3^-(aq) + 2\,H_2O(l) \rightarrow 2\,NO(g) + 3\,SO_4^{2-}(aq) + 4\,H^+(aq)$

(c) $2\,KBr(aq) + 2\,H_2SO_4(aq) \rightarrow K_2SO_4(aq) + Br_2(aq) + SO_2(g) + 2\,H_2O(l)$

(a) Step 1: Assign oxidation numbers to each element:

$$\overset{+2\,-2}{MnO} + \overset{+4\,-2}{PbO_2} \longrightarrow \overset{+7\,-2}{MnO_4^-} + \overset{+2}{Pb^{2+}}$$

Step 2: Diagram the electrons lost and the electrons gained per atom. Balance the electrons lost and gained.

$$\overbrace{\overset{+2\,-2}{MnO} + \overset{+4\,-2}{PbO_2} \longrightarrow \overset{+7\,-2}{MnO_4^-} + \overset{+2}{Pb^{2+}}}^{2 \times (5\,e^-\text{ lost})}_{5 \times (2\,e^-\text{ gained})}$$

Step 3: Determine what balancing coefficients are needed for the elements being oxidized and reduced.

$$\overbrace{\overset{+2\,-2}{2\,MnO} + \overset{+4\,-2}{5\,PbO_2} \longrightarrow \overset{+7\,-2}{2\,MnO_4^-} + \overset{+2}{5\,Pb^{2+}}}^{2 \times (5\,e^-\text{ lost})}_{5 \times (2\,e^-\text{ gained})}$$

Step 4: Balance oxygen and hydrogen as necessary.

$2\,MnO + 5\,PbO_2 + 8\,H^+ \longrightarrow 2\,MnO_4^- + 5\,Pb^{2+} + 4\,H_2O$

Step 5: Check the equation.

Step 6: Add symbols for physical states.

(b) Step 1: Assign oxidation numbers:

$$\overset{+4-2}{SO_2} + \overset{+5-2}{NO_3^-} \longrightarrow \overset{+2-2}{NO} + \overset{+6-2}{SO_4^{2-}}$$

Step 2: Diagram and balance electrons lost and gained.

$$\overbrace{\overset{+4-2}{SO_2} + \underbrace{\overset{+5-2}{NO_3^-} \longrightarrow \overset{+2-2}{NO}}_{2 \times (3\ e^-\ gained)} + \overset{+6-2}{SO_4^{2-}}}^{3 \times (2\ e^-\ lost)}$$

Step 3: Determine balancing coefficients:

$$\overbrace{3\,\overset{+4-2}{SO_2} + \underbrace{2\,\overset{+5-2}{NO_3^-} \longrightarrow 2\,\overset{+2-2}{NO}}_{2 \times (3\ e^-\ gained)} + 3\,\overset{+6-2}{SO_4^{2-}}}^{3 \times (2\ e^-\ lost)}$$

Step 4: Balance oxygen and hydrogen as necessary.

$$3\,SO_2 + 2\,NO_3^- + 2\,H_2O \longrightarrow 2\,NO + 3\,SO_4^{2-} + 4\,H^+$$

Steps 5 & 6: Check and add physical states.

(c) Steps 1, 2, and 3: Assign oxidation numbers, diagram and balance electrons lost and gained, and determine balancing coefficients:

$$\overbrace{2\,\overset{+1-1}{KBr} + \underbrace{\overset{+1+6-2}{H_2SO_4} \longrightarrow \overset{+1+6-2}{K_2SO_4} + \overset{0}{Br_2}}_{(2\ e^-\ gained)} + \overset{+4-2}{SO_2}}^{2 \times (1\ e^-\ lost)}$$

Step 4: Balance oxygen and hydrogen as needed. This equation appears somewhat confusing with K_2SO_4 and SO_2 both as products from H_2SO_4. According to the electrons gained, only one of the H_2SO_4 is oxidized. There must be another H_2SO_4 to provide the anion to balance the K^+ from KBr. Also, a source of oxygen for the SO_2 is needed. If a coefficient of

Chapter 17: Oxidation-Reduction Reactions   199

2 is placed in front of H₂SO₄ to provide the additional sulfate ion needed for K₂SO₄, it is seen that the components of 2 water molecules are present on the left side of the equation, and it can be balanced by adding 2 H₂O to the right side.

$$2\text{ KBr} + 2\text{ H}_2\text{SO}_4 \rightarrow \text{K}_2\text{SO}_4 + \text{Br}_2 + \text{SO}_2 + 2\text{ H}_2\text{O}$$

Steps 5 & 6: Check and add physical states.

17.31 (a) $3\text{ CuS(s)} + 2\text{ NO}_3^-\text{(aq)} + 8\text{ H}^+\text{(aq)} \rightarrow 3\text{ Cu}^{2+}\text{(aq)} + 3\text{ S(s)} + 2\text{ NO(g)} + 4\text{ H}_2\text{O(l)}$

(b) $\text{ClO}^-\text{(aq)} + 2\text{ I}^-\text{(aq)} + 2\text{ H}^+\text{(aq)} \rightarrow \text{Cl}^-\text{(aq)} + \text{I}_2\text{(aq)} + \text{H}_2\text{O(l)}$

(c) $6\text{ Zn(s)} + \text{As}_2\text{O}_3\text{(s)} + 12\text{ H}^+\text{(aq)} \rightarrow 6\text{ Zn}^{2+}\text{(aq)} + 2\text{ AsH}_3\text{(g)} + 3\text{ H}_2\text{O(l)}$

(a) Steps 1, 2 and 3: Assign oxidation numbers, diagram and balance electrons lost and gained, and determine balancing coefficients:

$$\overbrace{3\overset{+2-2}{\text{CuS}} + 2\overset{+5-2}{\text{NO}_3^-} \rightarrow 3\overset{+2}{\text{Cu}}{}^{2+} + 3\overset{0}{\text{S}} + 2\overset{+2-2}{\text{NO}}}^{3 \times (2\text{ e- lost})}$$

$$\underbrace{\phantom{3\text{ CuS} + 2\text{ NO}_3^- \rightarrow 3\text{ Cu}^{2+} + 3\text{ S} + 2\text{ NO}}}_{2 \times (3\text{ e- gained})}$$

Steps 4, 5, and 6: Balance oxygen and hydrogen as necessary, check, and add physical states.

$$3\text{ CuS} + 2\text{ NO}_3^- + 8\text{ H}^+ \rightarrow 3\text{ Cu}^{2+} + 3\text{ S} + 2\text{ NO} + 4\text{ H}_2\text{O}$$

(b) Steps 1, 2 and 3: Assign oxidation numbers, diagram and balance electrons lost and gained, and determine balancing coefficients:

$$\overbrace{\overset{+1-2}{\text{ClO}^-} + 2\overset{-1}{\text{I}^-} \rightarrow \overset{-1}{\text{Cl}^-} + \overset{0}{\text{I}_2}}^{(2\text{ e- gained})}$$

$$\underbrace{\phantom{\text{ClO}^- + 2\text{ I}^- \rightarrow \text{Cl}^- + \text{I}_2}}_{2 \times (1\text{ e- lost})}$$

Steps 4, 5, and 6: Balance oxygen and hydrogen as necessary, check, and add physical states.

$$\text{ClO}^- + 2\text{ I}^- + 2\text{ H}^+ \rightarrow \text{Cl}^- + \text{I}_2 + \text{H}_2\text{O}$$

(c) Steps 1, 2 and 3: Assign oxidation numbers, diagram and balance electrons lost and gained, and determine balancing coefficients:

$$\underbrace{6\ \overset{0}{Zn}\ +\ \overset{+3\ -2}{As_2O_3}\ \overset{\overbrace{\quad\quad\quad\quad}^{(2\ \times\ 6\ e^-\ gained)}}{\longrightarrow}\ 2\ \overset{-3\ +1}{AsH_3}\ +\ 6\ \overset{+2}{Zn^{2+}}}_{6\ \times\ (2\ e^-\ lost)}$$

Steps 4, 5, and 6: Balance oxygen and hydrogen as necessary, check, and add physical states.

$$6\ Zn\ +\ As_2O_3\ +\ 12\ H^+\ \rightarrow\ 2\ AsH_3\ +\ 6\ Zn^{2+}\ +\ 3\ H_2O$$

17.33 (a) $H_2O_2(aq) + I_2(aq) \rightarrow 2\ I^-(aq) + O_2(g) + 2\ H^+(aq)$; oxidation # of O in product is 0

(b) $H_2O_2(aq) + Sn^{2+}(aq) + 2\ H^+(aq) \rightarrow Sn^{4+}(aq) + 2\ H_2O(l)$; oxidation # of O in product is -2.

Once the missing product containing oxygen has been determined, each equation can be balanced by inspection. (a) Since the oxidation number of I is decreasing from 0 to -1, the oxidation number of O must increase. Therefore, H_2O_2 is oxidized to O_2. (b) Since the oxidation number of Sn is increasing from +2 to +4, the oxidation number of O must decrease. Therefore, H_2O_2 is reduced to H_2O.

17.35 (a) $Cr_2O_7^{2-}(aq) + CH_3OH(aq) + 8\ H^+(aq) \rightarrow 2\ Cr^{3+}(aq) + CO_2(g) + 6\ H_2O(l)$

(b) $3\ C_2H_4(g) + 2\ MnO_4^-(aq) + 2\ H_2O(l) + 2\ H^+(aq) \rightarrow 3\ C_2H_6O_2(aq) + 2\ MnO_2(s)$

(c) $5\ C_2H_4(g) + 12\ MnO_4^-(aq) + 36\ H^+(aq) \rightarrow 10\ CO_2(g) + 12\ Mn^{2+}(aq) + 28\ H_2O(l)$

(a) Steps 1, 2 and 3: Assign oxidation numbers, diagram and balance electrons lost and gained, and determine balancing coefficients:

$$\underset{+6\ -2}{Cr_2O_7^{2-}} + \underset{-2+1\ -2+1}{CH_3OH} \longrightarrow 2\ \underset{+3}{Cr^{3+}} + \underset{+4-2}{CO_2}$$

Electrons: 2 × (3 e⁻ gained) on Cr; (6 e⁻ lost) on CH₃OH → CO₂

Steps 4, 5, and 6: Balance oxygen and hydrogen as necessary, check, and add physical states.

$$Cr_2O_7^{2-} + CH_3OH + 8\ H^+ \rightarrow 2\ Cr^{3+} + CO_2 + 6\ H_2O$$

(b) Steps 1, 2 and 3: Assign oxidation numbers, diagram and balance electrons lost and gained, and determine balancing coefficients:

3 × (2 × 1 e⁻ lost)

$$3\ \underset{-2+1}{C_2H_4} + 2\ \underset{+7-2}{MnO_4^-} \longrightarrow 3\ \underset{-1+1\ -2}{C_2H_6O_2} + 2\ \underset{+4-2}{MnO_2}$$

2 × (3 e⁻ gained)

Steps 4, 5, and 6: Balance oxygen and hydrogen as necessary, check, and add physical states.

$$3\ C_2H_4 + 2\ MnO_4^- + 2\ H_2O + 2\ H^+ \rightarrow 3\ C_2H_6O_2 + 2\ MnO_2$$

(c) Steps 1, 2 and 3: Assign oxidation numbers, diagram and balance electrons lost and gained, and determine balancing coefficients:

5 × (2 × [6 e⁻ lost])

$$5\ \underset{-2+1}{C_2H_4} + 12\ \underset{+7-2}{MnO_4^-} \longrightarrow 10\ \underset{+4-2}{CO_2} + 12\ \underset{+2}{Mn^{2+}}$$

12 × (5 e⁻ gained)

Steps 4, 5, and 6: Balance oxygen and hydrogen as necessary, check, and add physical states.

$$5\ C_2H_4 + 12\ MnO_4^- + 36\ H^+ \rightarrow 10\ CO_2 + 12\ Mn^{2+} + 28\ H_2O$$

17.37 (a) $4\ HNO_3(aq) \rightarrow 4\ NO_2(g) + O_2(g) + 2\ H_2O(l)$

(b) $2\ Ce^{4+}(aq) + H_2O_2(aq) \rightarrow 2\ Ce^{3+}(aq) + O_2(g) + 2\ H^+(aq)$

(c) $H_2S(aq) + H_2O_2(aq) \rightarrow S(s) + 2\ H_2O(l)$

(a) Steps 1, 2 and 3: Assign oxidation numbers, diagram and balance electrons lost and gained, and determine balancing coefficients:

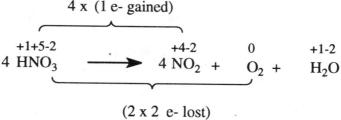

$$4 \text{ HNO}_3 \rightarrow 4 \text{ NO}_2 + \text{O}_2 + \text{H}_2\text{O}$$

with oxidation numbers: HNO$_3$ (+1, +5, −2), NO$_2$ (+4, −2), O$_2$ (0), H$_2$O (+1, −2); 4 × (1 e− gained) above, (2 × 2 e− lost) below.

Steps 4, 5, and 6: Balance oxygen and hydrogen as necessary, check, and add physical states.

$$4 \text{ HNO}_3 \rightarrow 4 \text{ NO}_2 + \text{O}_2 + 2 \text{ H}_2\text{O}$$

(b) Steps 1, 2 and 3: Assign oxidation numbers, diagram and balance electrons lost and gained, and determine balancing coefficients:

2 × (1 e− gained)

$$2 \text{ Ce}^{4+} + \text{H}_2\text{O}_2 \rightarrow 2 \text{ Ce}^{3+} + \text{O}_2$$

with oxidation numbers: Ce^{4+} (+4), H$_2$O$_2$ (+1, −1), Ce^{3+} (+3), O$_2$ (0); (2 × 1 e− lost) below.

Steps 4, 5, and 6: Balance oxygen and hydrogen as necessary, check, and add physical states.

$$2 \text{ Ce}^{4+} + \text{H}_2\text{O}_2 \rightarrow 2 \text{ Ce}^{3+} + \text{O}_2 + 2 \text{ H}^+$$

(c) Steps 1, 2 and 3: Assign oxidation numbers, diagram and balance electrons lost and gained, and determine balancing coefficients:

(2 e− lost)

$$\text{H}_2\text{S} + \text{H}_2\text{O}_2 \rightarrow \text{S} + 2 \text{ H}_2\text{O}$$

with oxidation numbers: H$_2$S (+1, −2), H$_2$O$_2$ (+1, −1), S (0), H$_2$O (+1, −2); 2 × (1 e− gained) below.

Steps 4, 5, and 6: Balance oxygen and hydrogen as necessary, check, and add physical states.

$$\text{H}_2\text{S} + \text{H}_2\text{O}_2 \rightarrow \text{S} + 2 \text{ H}_2\text{O}$$

17.39 The purpose of a salt bridge in the construction of a voltaic cell is to permit the flow of ions without complete mixing of the solutions.

17.41 (a)

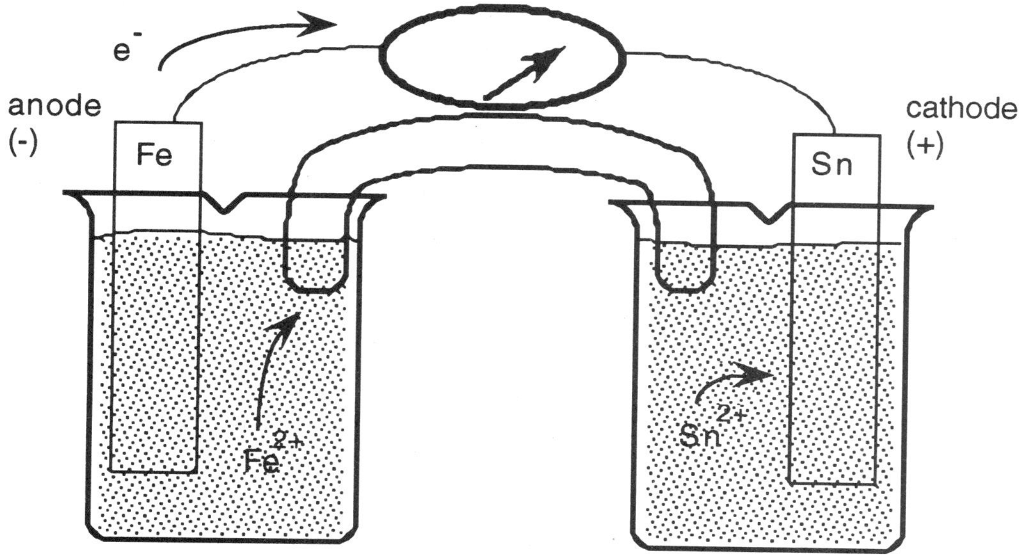

[Diagram of voltaic cell: anode (−) Fe electrode in left beaker with Fe^{2+} solution, cathode (+) Sn electrode in right beaker with Sn^{2+} solution, connected by salt bridge and external circuit with meter; electrons flow from Fe to Sn externally.]

(b) anode: $Fe(s) \rightarrow Fe^{2+}(aq) + 2e^-$ cathode: $Sn^{2+}(aq) + 2e^- \rightarrow Sn(s)$

17.43 (a) See Figure 17.7 in the text. In your diagram, the anode will be the electrode at which chlorine gas is produced. The cathode will be the electode at which sodium ion reacts to produce liquid sodium. The cathode will be negative and the anode will be positive. The current flows through the external circuit from the anode to the cathode. The sodium ions are attracted to the cathode, and the chloride ions are attracted to the anode.

(b) anode: $2\ Cl^-(l) \rightarrow Cl_2(g) + 2\ e^-$;   cathode: $Na^+(l) + e^- \rightarrow Na(l)$

electrolysis reaction: $2\ NaCl(l) \rightarrow 2\ Na(l) + Cl_2(g)$

17.45 (a, b)

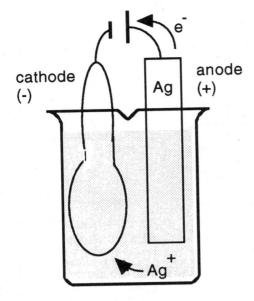

(c) anode: Ag(s) → Ag$^+$(aq) + e$^-$;   cathode: Ag$^+$(aq) + e$^-$ → Ag(s)

17.47 (a) element oxidized: (a) S, (b) S, (c) Br

(b) element reduced: (a) N, (b) I, (c) S

(c) oxidizing agent: (a) NO$_3^-$, (b) I$_2$, (c) H$_2$SO$_4$

(d) reducing agent: (a) HgS, (b) S$_2$O$_3^{2-}$, (c) Br$^-$

17.49 (a) 2 HNO$_2$(aq) + 2 I$^-$(aq) + 2 H$^+$(aq) → 2 NO(g) + I$_2$(aq) + 2 H$_2$O(l)

(b) 4 Ag(s) + 2 H$_2$S(aq) + O$_2$(aq) → 2 Ag$_2$S(s) + 2 H$_2$O(l)

(c) MoO$_3$(s) + 3 H$_2$(g) → Mo(s) + 3 H$_2$O(l)

(a) Using the half-reaction method:

$$2(HNO_2 + H^+ + e^- \rightarrow NO + H_2O)$$

$$\underline{2 I^- \rightarrow I_2 + 2 e^-}$$

$$2 HNO_2 + 2 I^- + 2 H^+ \rightarrow 2 NO + I_2 + 2 H_2O$$

(b) Using the half-reaction method:

$$2(2\,Ag + H_2S \rightarrow Ag_2S + 2\,H^+ + 2\,e^-)$$

$$\underline{O_2 + 4\,H^+ + 4\,e^- \rightarrow 2\,H_2O}$$

$$4\,Ag + 2\,H_2S + O_2 \rightarrow 2\,Ag_2S + 2\,H_2O$$

(c) Using the oxidation number method:

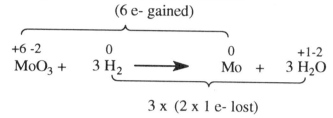

17.51 (a, b)

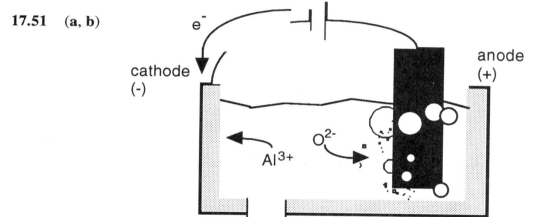

(c) $2\,Al_2O_3(l) \rightarrow 4\,Al(l) + 3\,O_2(g)$

Chapter 18:
Radioactivity and Nuclear Energy

18.1 Becquerel is credited with discovering radioactivity.

18.3 alpha particles, the nuclei of helium atoms with a mass of 4 and a charge of +2, limited penetrating power, $^{4}_{2}He$ (sometimes written $^{4}_{2}\alpha$); beta particles, electrons emitted from the nucleus, with lower energy than alpha particles but greater penetrating power, $^{0}_{-1}\beta$; gamma rays, massless radiation with great penetrating power, $^{0}_{0}\gamma$

18.5 Marie and Pierre Curie discovered polonium and radium.

18.7 Geiger-Müller counter

18.9 Activity is equal to the number of radioactive disintegrations per second.

18.11 A "one-curie source" is equivalent to a one gram sample of radium-226. It yields 3.7×10^{10} disintegrations per second.

18.13 The rem is a unit of radiation dosage that takes into consideration the difference in effectiveness of various types of radiation in causing damage. A rem is related to a rad by the following equation : dose (rems) = dose (rads) x BEF, where BEF is selected for the type of radiation.

18.15 A natural decay series is a series of nuclear transformations that occurs in nature and that converts a radioactive isotope into a stable nonradioactive isotope of a different element. The uranium-238 decay series goes no further than lead-206 because the nucleus of lead-206 is stable and therefore not radioactive.

18.17 (a) $^{228}_{89}Ac \rightarrow ^{0}_{-1}\beta + ^{228}_{90}Th$
(b) $^{3}_{1}H \rightarrow ^{0}_{-1}\beta + ^{3}_{2}He$

Since a beta particle has a mass of about 0, there is no change is mass number. However, the atomic number of the product nuclide must be greater by one, since electrical charge is conserved and the sum of the atomic numbers on the product side will equal the atomic number on the reactant side.

18.19 (a) $^{59}_{26}Fe \rightarrow ^{0}_{-1}\beta + ^{59}_{27}Co$
(b) $^{234}_{91}Pa \rightarrow ^{0}_{-1}\beta + ^{234}_{92}U$
(c) $^{60}_{27}Co \rightarrow ^{0}_{-1}\beta + ^{60}_{28}Ni$

18.21 $^{238}_{92}U \rightarrow 8\, ^{4}_{2}He + 6\, ^{0}_{-1}\beta + ^{?}_{?}?$

mass number = 238 - 8(4) - 6 (0) = 238 - 32 = 206
atomic number = 92 - 8(2) - 6(-1) = 92 -16 + 6 = 82 ∴ $^{206}_{82}Pb$ is the product

18.23 (a) $^{142}_{55}Cs \rightarrow {}^{142}_{56}Ba + {}^{0}_{-1}\beta$; beta decay
(b) $^{128}_{50}Sn \rightarrow {}^{128}_{51}Sb + {}^{0}_{-1}\beta$; beta decay
(c) $^{229}_{92}U \rightarrow {}^{225}_{90}Th + {}^{4}_{2}He$; alpha decay

18.25 1.56%

100% x 0.5 x 0.5 x 0.5 x 0.5 x 0.5 x 0.5 = 1.56%

Take 100% as the starting quantity and multiply it by 0.5 six times for six half-lives.

18.27 1.8 g

First, 1.0 mol of carbon-14 would have a mass of 14 grams. (two significant figures)
Next, $\frac{17\,200\;\cancel{yr}}{5730\;\cancel{yr}}$ = 3 half-lives.

Multiply 14 grams by 0.5 three times: 14 g x 0.5 x 0.5 x 0.5 = 1.8 g left

18.29 $^{14}_{7}N + {}^{1}_{0}n \rightarrow {}^{14}_{6}C + {}^{1}_{1}H$

18.31 The ratio of carbon-14 to carbon-12 remains constant in a growing tree because the tree is incorporating atmospheric carbon-14 into its structure as fast as the carbon-14 already in its structure is decaying. The ratio decreases steadily after the tree is cut down because the processes that would incorporate carbon-14 have stopped (e.g., growth), while the decay process continues.

18.33 starting nucleus is iron-58: $^{58}_{26}Fe + {}^{1}_{0}n \rightarrow {}^{59}_{26}Fe$

18.35 $^{241}_{95}\text{Am} + ^{4}_{2}\text{He} \rightarrow ^{243}_{97}\text{Bk} + 2\,^{1}_{0}\text{n}$

Write an incomplete nuclear equation, indicating what is unknown:
$$^{241}_{95}\text{Am} + ^{4}_{2}\text{He} \rightarrow ^{243}_{97}\text{Bk} + ?$$

The missing mass number is 2 and the missing atomic number is 0. The only particle with an atomic number of 0 is a neutron, so two neutrons must be formed.

18.37 bombarding particle is the nucleus of a carbon-12 atom: $^{246}_{96}\text{Cm} + ^{12}_{6}\text{C} \rightarrow ^{254}_{102}\text{No} + 4\,^{1}_{0}\text{n}$

Write an incomplete nuclear equation, indicating what is unknown:
$$^{246}_{96}\text{Cm} + ? \rightarrow ^{254}_{102}\text{No} + 4\,^{1}_{0}\text{n}$$

The missing mass number is 12 and the missing atomic number is 6. Carbon has atomic number 6, so the bombarding particle is the nucleus of a carbon-12 atom.

18.39 A critical condition for a fission process is the condition in which the number of fission events per unit time remains constant.

18.41 (a) $^{235}_{92}\text{U} + ^{1}_{0}\text{n} \rightarrow ^{90}_{37}\text{Rb} + ^{144}_{55}\text{Cs} + 2\,^{1}_{0}\text{n}$
 (b) $^{235}_{92}\text{U} + ^{1}_{0}\text{n} \rightarrow ^{90}_{38}\text{Sr} + ^{144}_{54}\text{Xe} + 2\,^{1}_{0}\text{n}$

(a) Write an incomplete nuclear equation, indicating what is unknown:
$$^{235}_{92}\text{U} + ^{1}_{0}\text{n} \rightarrow ^{90}_{37}\text{Rb} + ^{144}_{55}\text{Cs} + ?\,^{1}_{0}\text{n}$$

The mass number on the right must be increased by 2 for balance. Therefore, 2 neutrons are produced. (b) The same reasoning as in (a) above shows that 2 neutrons are produced.

18.43 See Figure 18.10 in the text. Be sure to include a core of radioactive material with control rods made of a moderating material, cooling fluid passing through and around the nuclear core, and a place for the now heated cooling fluid to exchange heat with water to make the steam to run a steam turbine.

18.45 An explosion cannot occur in a nuclear fission reactor because the uranium used in a nuclear reactor is only 3% uranium-235. This percentage of the fissionable isotope of uranium is too low for a nuclear explosion.

18.47 $0.137\,00$ amu $= 2.2750 \times 10^{-28}$ kg

Oxygen-16 has 8 protons and 8 neutrons. The total mass of nucleons is

mass of 8 protons =	8(1.007 28 amu) =	8.058 24 amu
mass of 8 neutrons =	8(1.008 66 amu) =	8.069 28 amu
total mass of nucleons:		16.127 52 amu
minus the nuclear mass:		-15.990 52 amu
mass defect (amu):		0.137 00 amu

mass defect (g) = $(\dfrac{1\text{ g}}{6.0221 \times 10^{23}\text{ amu}})(0.137\,00\text{ amu}) = 2.2750 \times 10^{-25}$ g

$= 2.2750 \times 10^{-28}$ kg

18.49 9.0×10^{-11} J

In order to calculate the energy in joules released, the total loss of mass in kg must be calculated:

mass (kg) = $(\dfrac{1\text{ kg}}{1000\text{ g}})(\dfrac{1\text{ g}}{6.0221 \times 10^{23}\text{ amu}})(\dfrac{0.0061\text{ amu}}{1\text{ nuclei}})(100\text{ nuclei}) = 1.0 \times 10^{-27}$ kg

$E = mc^2 = (1.0 \times 10^{-27}\text{ kg})(2.998 \times 10^8\text{ m/s})^2 = 9.0 \times 10^{-11}$ J

18.51 Nuclear fusion occurs naturally in stars such as the sun.

18.53 $^{2}_{1}H + ^{2}_{1}H \rightarrow ^{3}_{2}He + ^{1}_{0}n$

18.55 Nuclear fusion is more difficult to initiate than fission because extremely high temperatures are required for fusion. Fission has been used to provide the necessary energy to initiate fusion.

18.57 Radioactive iodine is used in two ways in medicine. It can be used for the diagnosis of thyroid problems and for the treatment of thyroid cancer. Different radioisotopes are used for the two purposes. In the diagnosis of thyroid problems, enough radioactive iodine is administered in a solution for a scan of the thyroid gland. In the treatment of thyroid cancer, radiation can be localized in the thyroid gland via radioactive iodine, and this radiation then kills the cancer cells.

18.59 A thickness gauge works by measuring the amount of radiation that passes through a plastic film. The amount of radiation that passes through the film is inversely related to the thickness of the film, i.e., the thicker the film, the less radiation passes through the film to reach a detector on the other side from the radioactive source.

Chapter 19:
Introduction to Organic Chemistry

19.1

$$H-\underset{\underset{H}{|}}{\overset{\overset{H}{|}}{C}}-C\equiv C-H \quad \quad \underset{H}{\overset{H}{>}}C=C=C\underset{H}{\overset{H}{<}}$$

109°, 180°, 120°

19.3

$$H-\underset{\underset{H}{|}}{\overset{\overset{H}{|}}{C}}-\underset{\underset{H}{|}}{\overset{\overset{H}{|}}{C}}-\underset{\underset{Cl}{|}}{\overset{\overset{H}{|}}{C}}-H \quad \text{1-chloropropane}$$

$$H-\underset{\underset{H}{|}}{\overset{\overset{H}{|}}{C}}-\underset{\underset{Cl}{|}}{\overset{\overset{H}{|}}{C}}-\underset{\underset{H}{|}}{\overset{\overset{H}{|}}{C}}-H \quad \text{2-chloropropane}$$

19.5

$CH_3CH_2CH=CH_2$ 1-butene

$\underset{CH_3}{\overset{H}{>}}C=C\underset{H}{\overset{CH_3}{<}}$ *trans*-2-butene

$\underset{H}{\overset{CH_3}{>}}C=C\underset{H}{\overset{CH_3}{<}}$ *cis*-2-butene

$CH_2=C\underset{CH_3}{\overset{CH_3}{<}}$ 2-methylpropene

19.7

$CH_3CH_2CH_2CH_2CH_2CH_3$

$CH_3CH_2CH_2CHCH_3$
    $|$
    CH_3

$CH_3CH_2CHCH_2CH_3$
       $|$
       CH_3

$CH_3CH_2\underset{\underset{CH_3}{|}}{\overset{\overset{CH_3}{|}}{C}}CH_3$

$CH_3\underset{\underset{CH_3}{|}}{\overset{\overset{CH_3}{|}}{CH}}CHCH_3$

19.9 (a) CH₃CHCH₂CH₂CH₂CH₃
          |
          CH₃

(b) CH₃CH₂CH₂CHCH₂CH₂CH₃
              |
              CH(CH₃)₂

(c) CH₃CH₂CHCH₂CH₂CH₃
          |
          CH₂CH₃

(d) CH₃CH₂CH₂CHCH₂CH₂CH₂CH₃
              |
              C(CH₃)₃

19.11 (a) CH₃CH₂CH₂CH₂CH₂CH₂CH₃

(b) CH₃C≡CCHCH₂CH₃
            |
            CH₂CH₃

(c) CH₃CHCCH₂CH₂CH₃ with CH₃ above and CH₃, CH₃ on the quaternary carbon

(d) 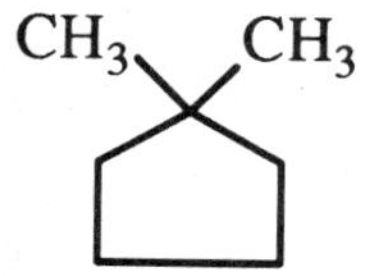 cyclopentane with two CH₃ groups on one carbon

19.13 (a) O=C=O   (b) CH₃CH₃   (c) CH₃CH₂CH₃

19.15 (a) CH₃C≡CH   (b) H₂   (c) ∼CH₂CCl₂(CH₂CCl₂)ₙ∼

19.17

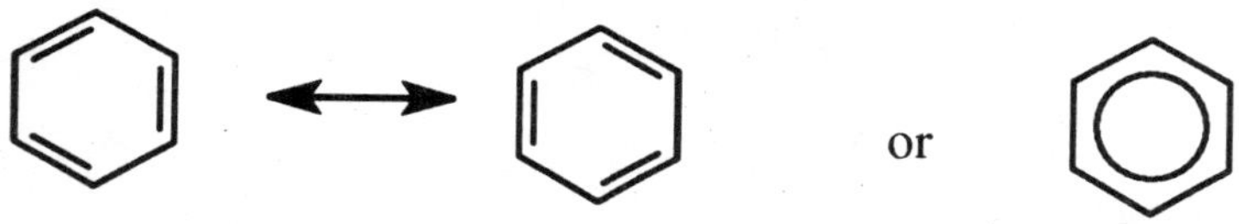

benzene ⟷ benzene   or   benzene with circle

The circle drawn inside a six carbon ring means six electrons spread out over the six carbon atoms by resonance.

19.19 The octane rating of a gasoline compares the antiknock performance of the gasoline to the performance of isooctane. Branched alkanes and aromatic hydrocarbons have the highest octane numbers.

19.21 highest octane number: (a) toluene

Chapter 19: Introduction to Organic Chemistry   215

19.23 (a) CH₂CH₂OH (b) CH₃CHCCH₃ (c) C₆H₅—CH₂—OH (d) CH₃CCH₂CH₂CH₃
            | | | |
            Cl   CH₃ CH₃   CH₃
                           |
                          OH

With (b) showing (CH₃)₂ on top carbon and OH below, and (d) showing OH on top and CH₃ below.

19.23 (a) $CH_2(Cl)CH_2OH$ (b) $CH_3CH(OH)C(CH_3)_2CH_3$ (c) C_6H_5—CH_2OH (d) $CH_3C(OH)(CH_3)CH_2CH_2CH_3$

19.25 (a) C_6H_5—OCH_2CH_3 (b) $CH_3CH(CH_3)CH_2CH_2OH$ (c) $CH_3C(OH)(CH_3)$-cyclopentyl (d) $CH_3CH(CH_3)CH_2OH$

19.27 (a) 1,2-ethanediol (ethylene glycol) (b) diphenyl ether (c) cyclopentanol

19.29 (a) $CH_3CH(CH_3)COCH_2CH_3$ (b) cyclopentanone (c) CH_3CH_2CHO (d) $HCHO$

19.31
(a) $CH_3CH(OH)CH_2CH_2CH_3 \xrightarrow{\text{oxidation}} CH_3COCH_2CH_2CH_3$

(b) $CH_3COCH_2CH_3 + NaBH_4 \longrightarrow CH_3CH(OH)CH_2CH_3$

(c) $CH_3CH_2CH_2CHO + CrO_3 \longrightarrow CH_3CH_2CH_2COOH$

19.33
(a) $CH_3CH_2CH(OCH_2CH_3)_2$ (the carbon bears H, OCH₂CH₃ above, OCH₂CH₃ below)

(b) $CH_3C(OCH_3)(OCH_3)(CH_3)$

19.35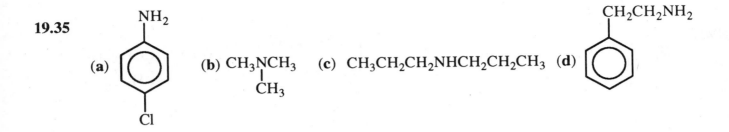

19.37 (pyridine structure) A heterocyclic compound has a noncarbon atom as part of a ring.

19.39 (a) $(CH_3)_2NH$ (b) CH_3I (c) NH_3

19.41 (a) $CH_3CH_2CH_2\overset{\overset{O}{\|}}{C}OCH_2CH_2CH_2CH_3$ (c) $CH_3\overset{\overset{O}{\|}}{C}OCH(CH_3)_2$

(b) 2-bromobenzoic acid (—CO_2H with ortho Br)   (d) $CH_3CH_2CH_2CH_2CH_2\overset{\overset{O}{\|}}{C}NHCH_3$

19.43 $CH_3CH_2CH_2CO_2H + NaOH \rightarrow CH_3CH_2CH_2CO_2^-Na^+ + H_2O$

19.45 $HCO_2CH_2CH(CH_3)_2 + NaOH \rightarrow HCO_2^-Na^+ + HOCH_2CH(CH_3)_2$

19.47
$$CH_3CH_2O-\underset{\underset{OCH_2CH_3}{|}}{\overset{\overset{O}{\|}}{P}}-OCH_2CH_3$$

19.49
$$\sim\!\!\left(NHCH_2CH_2CH_2CH_2CH_2CH_2NH\overset{\overset{O}{\|}}{C}CH_2CH_2CH_2CH_2\overset{\overset{O}{\|}}{C}\right)_{\!n}\!\!\sim$$

Chapter 19: Introduction to Organic Chemistry   217

19.51 unsaturated: (b) 1-butene and (d) 2-butyne

Unsaturated hydrocarbons (alkenes and alkynes) contain carbon-carbon double or triple bonds.

19.53

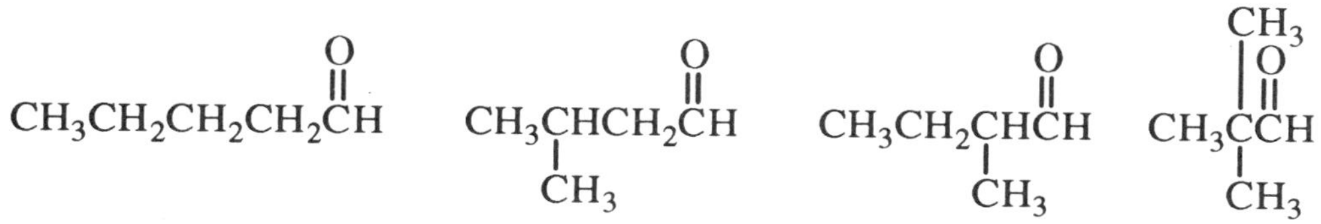

$$CH_3CH_2CH_2CH_2\overset{\overset{O}{\|}}{C}H \qquad CH_3\underset{\underset{CH_3}{|}}{C}HCH_2\overset{\overset{O}{\|}}{C}H \qquad CH_3CH_2\underset{\underset{CH_3}{|}}{C}H\overset{\overset{O}{\|}}{C}H \qquad CH_3\underset{\underset{CH_3}{|}}{\overset{\overset{CH_3}{|}}{C}}\overset{\overset{O}{\|}}{C}H$$

19.55

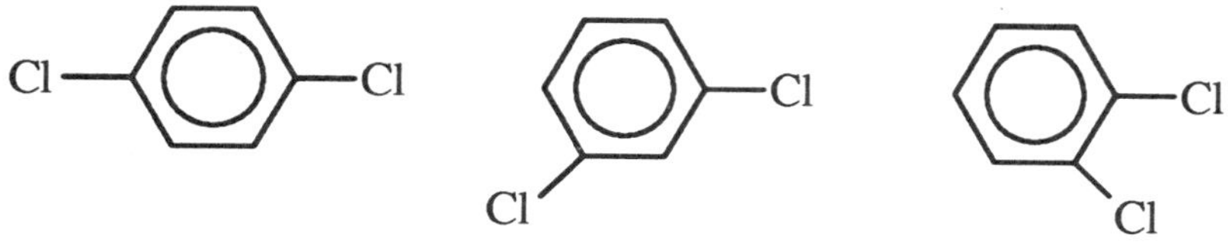

Cl—⌬—Cl   ⌬(Cl, Cl meta)   ⌬(Cl, Cl ortho)

19.57

$$CH_3CH_2CH_2CH_2OH \qquad CH_3\underset{\underset{CH_3}{|}}{C}HCH_2OH \qquad CH_3CH_2\underset{\underset{OH}{|}}{C}HCH_3 \qquad CH_3\underset{\underset{CH_3}{|}}{\overset{\overset{CH_3}{|}}{C}}OH$$

1-butanol   3-methyl-1-propanol   2-butanol   2-methyl-2-propanol

19.59

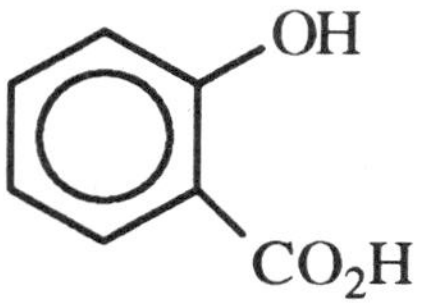

⌬(OH, CO₂H)   2-hydroxybenzoic acid   CH₃OH   methanol

19.61   HC≡CCH₂CH₃   CH₃C≡CCH₃   CH₂=C=CHCH₃   CH₂=CHCH=CH₂

19.63

$$\text{HCOH} + \text{HOCH}_2\text{CH}_3 \xrightarrow{\text{H}^+} \text{HCOCH}_2\text{CH}_3 + \text{H}_2\text{O}$$
(formic acid + ethanol → ethyl formate + water)

$$\text{CH}_3\text{COH} + \text{HOCH}_3 \xrightarrow{\text{H}^+} \text{CH}_3\text{COCH}_3 + \text{H}_2\text{O}$$
(acetic acid + methanol → methyl acetate + water)

19.65

H—O—C(=O)—C(=O)—O—H

19.67 CH₃CH₂CO₂CH₂CH₂CH₂CH₂CH₃

19.69
CH₃CH(NH₂)CO₂H

Chapter 20:
Biochemistry: The Chemistry of Life

20.1 Fibrous proteins are rod-shaped whereas globular proteins are more spherical in shape.

20.3 The R groups in amino acids can be polar or nonpolar. Polar R groups increase the compatibility of an amino acid with water, since water is polar. Nonpolar R groups decrease the compatibility of an amino acid with water.

20.5

20.7 120

The number of possible structures is 5! or 5 x 4 x 3 x 2 x 1 = 120.

20.9 The pleated sheet form of protein secondary structure involves long strands of protein running parallel with hydrogen bonding between the side-by-side chains. The result is a sheet-like structure that is pleated along the lines of the protein chains. Silk and some muscle fiber contain this structure.

20.11 Tertiary structure of a protein refers to the overall shape of a protein that results from the protein chain folding back on itself.

20.13 Hemoglobin is an example of a protein that has quaternary structure. Quaternary structure is the result of combining two or more proteins in order to perform a specific function, such as the transport of oxygen by hemoglobin.

20.15 enzyme

20.17 substrate

20.19 carbonyl groups (aldehydes and ketones) and hydroxy groups

20.21 trisaccharide

20.23 Glucose is the building block. Starch and cellulose differ in the position of the -OH group on C-1. In starch, this -OH group is down (α) and in cellulose it is up (β).

20.25 The structure of glycogen facilitates the release of glucose by having many branches. Glucose is easily released only from the ends of chains. Therefore the more ends, the more rapidly glucose can be released in the body upon demand.

20.27 two components of a typical triglyceride: glycerol and fatty acids

$$\begin{array}{c} H_2C-O-\overset{\overset{O}{\|}}{C}-R \\ HC-O-\overset{\overset{O}{\|}}{C}-R' \\ H_2C-O-\overset{\overset{O}{\|}}{C}-R'' \end{array}$$

20.29 A beneficial use for cholesterol in the body is to serve as the basis for the formation of other steroids, e.g., sex hormones. Cholesterol can be harmful to the body by the formation of plaque which deposits in arteries, and by precipitating as gall stones.

20.31 a purine   a pyrimidine

20.33

20.35 The DNA double helix is held together by hydrogen bonding between complementary bases on the antiparallel strands of the double helix.

20.37 gene